THOMAS MOORE

BRITISH FERNS

AND

THEIR ALLIES

Elibron Classics
www.elibron.com

Elibron Classics series.

ISBN 1-4212-7081-1 (paperback)
ISBN 1-4212-7080-3 (hardcover)

This Elibron Classics Replica Edition is an unabridged facsimile of the edition published in 1862 by Routledge, Warne, and Routledge, London.

BRITISH FERNS

AND

THEIR ALLIES:

AN ABRIDGMENT OF

THE "POPULAR HISTORY OF BRITISH FERNS,"

AND COMPRISING THE

Ferns, Club-Mosses, Pepperworts, & Horsetails.

BY

THOMAS MOORE, F.L.S., F.H.S., &c.

AUTHOR OF "THE HANDBOOK OF BRITISH FERNS,"
"THE FERNS OF GREAT BRITAIN AND IRELAND, NATURE PRINTED,"
"THE OCTAVO NATURE-PRINTED BRITISH FERNS," ETC.

Illustrated by W. S. Coleman.

NEW EDITION.

LONDON:
ROUTLEDGE, WARNE, AND ROUTLEDGE,
2, FARRINGDON STREET.
NEW YORK: 56, WALKER STREET.
1862.

"British Ferns."

Savill and Edwards, Printers, Chandos Street.

PREFACE.

This little volume, which is virtually an abridgment of the *Popular History of British Ferns*, having been provided for the use of beginners in the study of these charming wild plants, it has been made an especial object to avoid technicalities as far as was possible, and to afford plain, and, it is to be hoped, easily understood, descriptions of the plants.

The varieties which are now known to occur among our wild Ferns, are so numerous, that only a few of the more striking have been noticed. A full enumeration of them would have taken up much space, and would only have distracted the attention of the learner from the species which it is important he should first become acquainted, with.

Being intended as a first book, many details of structure, of distribution, &c., have also been omitted with the view of simplifying the main features of the subject. When these have been well mastered, the inquirer will meet with additional useful information in the more extended *Popular History of British Ferns*,* already mentioned.

As a help to those readers who may not be familiar with the scientific names derived from the Greek and Latin languages, these names have all been accentuated in the Index.

* A POPULAR HISTORY OF BRITISH FERNS, and the Allied Plants. By THOMAS MOORE, F.L.S., F.H.S. 3rd and revised edition. London: Routledge, Warne, and Routledge, Farringdon Street.

BRITISH FERNS.

CHAPTER I.

POPULARITY OF FERNS—STRUCTURE—STEMS: CAUDEX, RHIZOME—FRONDS—VENATION—FRUCTIFICATION.

THE native Ferns of Great Britain have undoubtedly become attractive objects of inquiry and of study, amongst those who take interest in natural objects. Among the many causes which may have conduced to bring about this result, probably the most active is this, that these plants are for the most part objects of exquisite elegance. This fact is, indeed, apparent, whether they are superficially examined as to their external appearance, or whether they are investigated with the view to analyse their minute structure. Then, again, they have become fashionable; and they are, moreover, neither very numerous, nor very inaccessible, and consequently their study opens a field which even those who have not much leisure may hope to compass, and for which the greater part, at least, of the materials may be obtained without much difficulty. To these inducements, it may be added, that they are plants for the most part very easily cultivated, and of all others perhaps the best adapted to parlour or window culture. Hence it follows that, besides the interest that may arise in the collection and preservation of the British Ferns in the herbarium, and in the study of them in the dried state, there is to be added the pleasure to be derived from their cultivation, and the opportunities thereby afforded of studying and admiring them in the living state. Those who desire a thorough knowledge of them should certainly, if possible, adopt the latter method of study, as it reveals many curious and interesting features which are not to be learned from the most patient investigations, which are carried on by the aid of the dried plants alone. Notwithstanding, however, the advantage of studying the plants in the living state, it is to be borne in mind, that all the essential points necessary for the recognition of the species may be present and available in well-selected herbarium specimens, so that those who have not convenience for cultivating them, may yet store up in their cabinets ample materials for amusement and instruction in leisure hours.

There is something peculiarly fascinating in the graceful outline and disposition of parts so common among the Ferns. Gay colours are indeed generally wanting to them, and they wear while in life a livery of sober green, which can scarcely be said to gain ornament from the brownish scales, with which in some of our native species it is associated. In some of the Ferns of tropical countries, however, as for example in some species of *Gymnogramma* and *Cheilanthes*, the lower surface is covered more or less with a silvery or golden powder, which adds considerably to their beauty; and there is among exotic Ferns considerable variety, even of the tints of green. The more sober-tinted natives of our northern latitude have, however, but comparatively little of such variety of hue; so that it is not in the colouring that their attractions rest, nor is it in their endurance, for a large proportion of the native species lose all their beauty as soon as the frost reaches them, and for nearly one-half of the year are dormant unless artificially sheltered. We may conclude, then, that it is the elegant forms and graceful habits of the majority of the Ferns, native and exotic, which render them so generally pleasing, even to those who are slow to perceive beauty apart from rich and gaudy colouring.

The number of species of British Ferns may be taken at from forty to fifty, according as some of the more doubtful forms are ranked as species or varieties. There are, however, a much larger number of curious and interesting variations.

What is a Fern? This question will be best answered by means of a familiar comparison.

Every reader of this book, aided by that intuitive perception which has grown up with the growing faculties and acquired strength from the little experiences of childhood and youth, will know what is meant by a flower. We must take for granted, that all those into whose hands these pages may fall, are familiar with such natural productions as the Buttercup, the Poppy, the Brier-rose, the Daisy, the Dandelion, and others, so profusely dispersed over the meadows and corn-fields, along the hedgerows, and by the waysides;

even the young ears of corn and the spikes of the meadow grasses must be well-remembered objects. Now, these all afford examples of flowers, either separate or collected into groups of varied form and character. The Daisy and Dandelion heads have been often plucked to be made into floral chains, and the Buttercups, the Roses, and various other flowers have been as often gathered for the rural garland. The plants which bore these brilliantly coloured parts which the tiny fingers chiefly desired to gather, bore other parts which were mostly green; in these latter the same intuitive perception learned to recognise the leaves. Now, besides the stem which bears them all, these two kind of "organs," as they are called—the leaves and the flowers—are the principal conspicuous parts of plants.

What, then, is a Fern? A Fern may be described, in a popular way, as a plant which bears leaves only, and no flowers. Still there remains the difficulty of how to distinguish a Fern which never bears flowers, from some other plant which does bear flowers, but from which they are temporarily absent. A little patient and attentive observation will overcome this seeming, and to the beginner real, difficulty. The course to be taken is this:—Search for what seems to be a full-grown plant: it will rarely happen that young plants, not in the fertile state to be presently mentioned, will occur without mature ones in the vicinity; examine the under surface of the leaves, and brown dust-like patches, round or elongated or in lines, will be seen placed here and there, and generally arranged with much regularity. These patches are in reality heaps—vast accumulations—of the minute seeds.

Now, as the leaves of those plants which NEVER bear flowers, bear these dust-like patches of seeds or spores, as they are technically termed, it is on their presence that the novice must depend for the *assurance* that his plant is a Fern. Imperfect as this test may be, and unsatisfactory as it doubtless is to the advanced student, there is in truth no other available guide-mark at the starting-point, nor until the eye has become familiarized with the peculiar appearances by aid of which Ferns may be recognised at first sight. The ready recognition of a Fern from other plants at first sight must be the result of experience gained by observation, towards the acquirement of which good and characteristic figures are valuable helps.

Ferns, then, are flowerless plants. They are furnished with roots, by which they obtain nourishment from the soil.

They have stems, by which their conspicuous parts are borne up and supported. They have leaves, or fronds, to which their elegance is due; and these leaves bear on some part of their surface, but usually on the lower face, the seeds by which the plants may be propagated. These are the several parts or external organs of the plants.

The proper roots of Ferns are entirely fibrous, and they proceed from the under side of those stems, which assume the prostrate or creeping mode of growth; but when the stem grows erect, they are produced towards its lower end on all sides indifferently, and proceed from among the bases of the decayed leaves or fronds. Fibrous roots are so called from their consisting of little thread-like parts or fibres; these, as they extend by growth at their points, insinuate themselves into the earth, so that in process of time it becomes filled with their ramifications. They often form entangled masses, but are not always sufficiently numerous for this. The fibres of Ferns are mostly of a rigid or wiry texture; and in the younger portions are often more or less covered with fine, soft, downy hairs, which become lost as they get older. It is by means of these organs chiefly that Ferns, and all the more highly developed races of plants, are nourished.

The stem of a Fern forms either an upright stock, called a caudex, which in our native species seldom elevates itself above the surface of the ground, but in certain exotic ferns reaches from thirty to fifty feet or more in height, and gives a tree-like character to the species; or it extends horizontally either on or beneath the surface of the soil, and forms what is called a rhizome or creeping stem. These creeping stems, when not buried in the earth, are generally clothed with hairs or scales, and sometimes to such an extent as to become quite shaggy; they vary greatly in size, some being as thick as one's wrist, and others, as in our native *Hymenophyllums*, as fine as threads.

The common Polypody has the thickest rhizome of any of the creeping British species: in this it is nearly as thick as one's thumb; but that of the common Bracken, or *Pteris*, which is formed under ground, creeps the most extensively. The *Osmunda*, or Flowering Fern, as it is called, is, of the native upright-growing species, that which most readily gains height; and very old plants of this may sometimes be found with bare stems of a foot or more in length. The stems of the common Male Fern, of *Lastrea montana*, and of *Polystichum angulare*, have also a tendency, though ina less degree, to this upright mode of growth; but it never becomes apparent, except in the case of very aged plants.

The leaves of Ferns are generally called fronds, and as this latter term is much the more appropriate, we shall henceforward adopt it, with this general explanation, that it means the leaf-like organs which are borne on the proper stem. The leaf-like character they bear has led some persons to reject the term *frond* altogether, and to consider them as true leaves; but since they grow by development from their apex, which botanists say leaves do not, and since they produce, from some part of their surface, what in their case stands in the place of flowers, there is no more reason why they should be called leaves, than should the leaf-like stems of *Cactuses*, or of the curious hot-house plants called *Xylophylla*, each of which afford examples of plants bearing flowers on what appear to be leaves, but which are in reality stems. The frond or leafy part of a Fern is, however, not to be classed among stems; and hence, since it is of intermediate character between a leaf and a stem, the distinctive name of *frond* seems to be conveniently applied to it.

As there are no proper flowers produced by the Ferns, it is in the fronds that we must seek for that ornamental aspect which renders them such general favourites. The fronds alone, however, afford almost endless variety. In some cases they are very large, in others very small; some are quite simple and not at all divided, others are divided almost beyond computation into little portions or segments, and it is these much-divided fronds which, generally speaking, are the most elegant.

Even in the few species which are natives of Britain, this variety of size and form is very obvious, some kinds not being more than two or three inches, while others are from five to six feet or more in height; some are quite simple, and others are cut into innumerable small segments. There is much variety of texture too; some being thin and delicate, almost transparent, others thick and leathery, and some perfectly rigid. There is again variety of colour; some are pale green, some are deep green, some are blue-green, some dark brownish, scarcely green at all; some are smooth and shining, others opaque; and some few are covered with hair-like scales.

The duration of the fronds of many species is comparatively short; they come up in spring, and in some cases the earliest of them do not last till autumn; in others they continue until touched by frost, from which the more robust of them shrink, even as the tender sorts do from drought as well as frost. Others are much more durable, and the plants, if in a moderately sheltered situation, become evergreen. These latter should be most extensively adopted for culture where ornamental effect is an object.

The fronds of Ferns consist of two parts—the leafy portion; and the stalk, which latter is called the *stipes.* The continuation of the stalk, in the form of a rib extending through the leafy portion, and becoming branched when the frond is divided, is called the *rachis.* If the frond is compound, that is, divided, so that there is another set of ribs besides the principal one, the latter is called the primary rachis, and the former the secondary rachis. In but few cases are our native species more highly compound than this. In practice, when the outline or division of the frond is mentioned, it is generally the leafy portion only that is referred to, exclusive of the stipes.

The stipes is generally furnished more or less with membranous scales, which are sometimes few and confined to the base, and at other times continued along the rachis. Sometimes these scales, which are generally brown, are large and so numerous that the parts on which they are situated acquire a shaggy appearance. The form of the scales, as well as their number and position and even colour, is found to be tolerably constant in the different species or varieties, and hence they sometimes afford marks of recognition. Whenever they are produced along the rachis, as well as on the stipes, they are invariably largest at the base, and become gradually smaller upwards.

In some species the leafy portion of the frond is undivided, that is to say, the margins are not scalloped or cut away at all: an example of this occurs in the common Hart's-tongue. More commonly, however, the margin is more or less divided.

The simplest mode of division is that where the margin of the frond is deeply divided or scalloped out at short intervals, the incision extending inwards nearly to the rachis, but not reaching it: this slightly divided form is called *pinnatifid.*

The fronds are sometimes divided down to the rachis, which is, as it were, quite bared of the contiguous leafy expanded portion, and when this occurs, the frond is said to be pinnate; and in this case, each of the distinct leaf-like divisions is called a *pinna.* When these pinnæ are divided again upon precisely the same plan, the frond becomes *bipinnate,* or twice pinnate; but if the pinnæ are only deeply lobed *they*, like the frond when similarly divided, are said to be pinnatifid.

When the fronds are thrice pinnate, they are called *tri-pinnate;* and in all other more intricate forms, they are called *decompound;* but this seldom occurs in our native kinds, the nearest approach to it being in very vigorous plants of the common Bracken, and in some of the *Lastreas*, when very largely developed.

The young fronds of the Ferns before being developed are arranged in a very curious manner, the rachis being rolled inwards volutely from the point to the base. In the compound sorts the divisions are each again rolled up in a similar way. This arrangement is what is called *circinate.* All the British species, with two exceptions, are folded up in this way, so that their development consists of an unrolling of the parts of the frond. The exceptions mentioned, are the Moonwort and the Adder's-tongue, in both of which the fronds in the undeveloped state are folded straight.

The substance of the fronds is traversed by veins variously arranged; in some species forming straight, nearly parallel lines, in others joined together like network. The manner in which the veins are disposed is called the *venation*, and the nature of this venation affords useful data in the division of the Ferns into family groups. It is from some part of these veins, determinate in position, that the clusters of fructification proceed, the part to which they are attached being called the *receptacle.*

A correct appreciation of the nature and position of the receptacle with reference to the veins, is of considerable importance in the study of the family groups and the individual kinds. In some, though few of the native kinds, it is projected beyond the margin, and the little cases of seeds are collected around the extremity of the vein, which projects outwards from the edge of the leaf. More commonly, however, the veins stop within the margins, and the seed-cases grow in round or elongated clusters, sometimes placed at their ends, sometimes along their sides, appearing on the lower surface of the fronds.

No flowers are produced, but the plants bear, generally, great abundance of seed-like bodies, which are technically called *spores*, and are contained in little hollow cases of very singular construction. Collectively, these cases and their contents are called the *fructification;* and the cases themselves are attached in the different species to certain determinate portions of the veins, which are thickened and form the *receptacles.* Each separate mass or cluster of the seed-cases is called a *sorus*, but being generally spoken of collectively, the plural term *sori* becomes much more frequently used. The sori are *marginal* when they project beyond the margin, and *dorsal* when occupying some part of the under surface of the frond.

The seed-cases—called also *spore-cases*, or *sporangia*, or *thecæ*—are mostly minute roundish-oval bodies, containing one cavity, and nearly surrounded by a jointed vertical band called a *ring*, which is continued from the base so as to form a short stalk, by which they are attached. When they have reached maturity, the elasticity of the ring bursts the case by an irregular transverse fissure, and the seeds or spores, in the shape of fine dust, which is almost invisible, become dispersed. This occurs in the majority of the native species. In *Trichomanes* and *Hymenophyllum*, however, the elastic band is horizontal or oblique, not vertical; and in *Osmunda*, *Botrychium* and *Ophioglossum*, the spore-cases are two-valved, and destitute of the elastic ring.

In a considerable proportion of the known species of Ferns, and in the majority of those which are natives of Great Britain, the sori are covered in the earlier stages of growth by what is commonly called the *indusium.* This is mostly a thin transparent membranous scale of the same general form as the sorus itself, at first completely covering or enclosing the young seed-cases, but eventually, by their growth, becoming disrupted at its margins and thrust back, or not unfrequently cast off before the maturity of the seeds. Some species, however, never bear any visible indusium, even in their earliest stages. The presence or absence of an indusium, or cover to the seed-heaps, is consequently one of the technical points by which Ferns are divided into groups of moderate extent.

In some Ferns the indusium, or cover, or at least what is considered analogous to it, is cup-shaped, containing the seed-cases; but this form is of very rare occurrence among the native species, and exists only in *Trichomanes* and the *Hymenophyllums.*

Taking now a retrospective glance, we have seen that the Ferns are, as regards external structure, flowerless plants, having erect or creeping stems, which bear the leaf-like fronds; and on some part of the surface of the fronds, usually the lower side, but sometimes the margin, are borne the clusters of seeds, which, in the majority of the native species, are, when young, furnished with a membranous scale-like cover.

The subject of internal structure we may here pass over with the remark that the Ferns belong to the lowest group of vegetation, which is especially remarkable for its loose and often succulent texture, owing to the absence, or the paucity, of those tissues which give firmness and elasticity to the higher orders of plants. The Ferns, however, are the highest members of this lower group, and hence we find them possessing, to some extent, both woody and vascular tissue, mixed up with the more succulent cellular tissue What these tissues are, may be found explained in any elementary book on physiological botany.

CHAPTER II.

PROPAGATION AND CULTURE—TOPOGRAPHICAL ASPECT—USES—PRESERVATION FOR THE HERBARIUM.

NATURALLY Ferns are propagated by means of the spores. These spores, which are somewhat analogous to seeds, being like them endowed with that mystery—the vital germ, when placed under fitting conditions, become developed into young plants; but they differ from seeds in some important particulars.

All true seeds have a determinate structure. They have an embryo, provided with special organs; there is the plumule, or germ of the descending axis, the origin of the stem, and there is the radicle, or germ of the descending axis, the origin of the root. When a seed is planted, in whatever position it may chance to have been deposited in the soil, the young root or radicle strikes downwards, and the young stem or plumule grows upwards.

The Fern spores have none of these determinate parts, but are, as it were, homogeneous atoms; and when placed under circumstances which induce germination, that part which lies downwards produces the root, and that part which lies upwards produces the rudimentary stem. The spores themselves are minute vesicles of cellular tissue. As they grow, this vesicle becomes divided into others, which again multiply and enlarge, until they form a minute green leaf like primordial scale or germ-frond, technically called the *prothallus*. From this the axis with its roots and stem are eventually developed.

In annual Ferns the mature character is soon attained, but in others, two or more years of growth is required before they reach maturity, and a much longer period is of course necessary to the maturity of those which acquire tree-like stems. They, however, in most cases, soon begin to assume something of their peculiar appearance.

In these minute and almost invisible atoms, no less than in the more ponderous materials which surround us, we discover the impress of Almighty and Creative power. They teem with life! No commixture of elementary matter—no electric shock guided by human agency, can originate that vitality. Truly, the hand that made them is Divine!

The requisite conditions to induce the germination of the spores of Ferns, in addition to the degree of heat proper for the particular species to which they belong, is simply contact with a continually damp earthy surface. Diffused light is favourable to the young growth as soon as it begins to form. It matters little in what way the principal conditions are supplied. In confined situations congenial to Ferns, the spores, which are shed as soon as they reach a certain degree of ripeness, germinate freely on any undisturbed surface with which they come in contact, whether it be the damp soil, or damp brickwork, or the sides of the pots in which the plants are growing. They grow very readily on the rough surface of a piece of sandstone rock just kept moistened.

The most convenient way, however, to raise Ferns from the spores, where cultivation is the object, is to sow them on the surface of moist loamy soil, in pots of convenient size, the surface of the soil being kept an inch or more below the level of the pot rim, so that a piece of flat glass may be laid over the top, to secure a close and constantly moist atmosphere, and prevent rapid evaporation. The pots should be set in pans or feeders, in which water should be kept, so that the soil may be constantly damp. The spores are to be thinly scattered over the rough surface of the soil, and then kept covered with the glass.

A simple and convenient contrivance for sowing the spores, by which the progress of germination might be very readily watched, would consist in inverting a porous flowerpot in a shallow dish or pan of water, large enough to take also the rim of an enclosing bell-glass, which should cover some surface of the water. A small cup or vase, set on the top of the inverted pot, with two or three worsted siphons, would keep its sides always damp; the spores scattered over the sides of this moistened porous earthenware would find a proper nidus for their development, which might thus be watched with great facility. It is to be borne in mind, however, that the seedling plants are not so readily transplanted from an earthenware or stone surface, as they are when growing on the soil.

The general features of culture—which it will be sufficient here to notice—are shade, shelter, and abundance of moisture. Neither of these are, however, essential to all the species; but when judiciously combined, they produce conditions under which all the species admit of being very successfully grown.

In the garden, Ferns seem only appropriately introduced on what is called rockwork, which generally means a bank of earth irregularly terraced with misshapen blocks of stone, or by masses of some other hard porous material, the vitrified conglomerations formed in the burning of bricks being that most commonly employed. With taste in the distribution of these and such-like materials, and in the planting of the Ferns, a very pleasing effect may be produced; and on rockwork of this kind, if it be erected in a shaded and sheltered situation, and liberally supplied with percolating (not stagnant) water, and if the soil be of a texture which will admit of being thus constantly moist without becoming soddened, nearly all the English Ferns may be grown successfully. The most sunny, most exposed, and least moistened positions on the rockwork should be appropriated to those species which grow naturally in situations to which these conditions afford the nearest resemblance; while, on the other hand, the kinds which naturally prefer the deepest shade and the dampest soil, should be placed in the positions where these conditions are most nearly imitated.

The most interesting mode for the amateur Fern-grower consists in the cultivation of the plants under glass, either in pots, or planted in a Wardian case. All the species admit of being grown in pots, and when developed under the protection of a covering of glass, acquire more than their natural delicacy of appearance. For the hardy Ferns, the frame or case in which they are grown should have a northern aspect. The plants must be kept cool in summer, by shading, by sprinkling, and by removing all impediments to a free circulation of air at night, not quite closing the frame even by day.

Wardian cases for Ferns, in which they may be planted out on rockwork, may be either of the size and nature of a small detached greenhouse, or of those window or balcony greenhouses made by enclosing within a projecting sash, a greater or smaller area external to the window; or they may be of smaller size and more finished workmanship, for the interior of dwelling-rooms, for staircase landings, or any other situation within doors where they can be moderately lighted.

The most proper soil for Ferns grown in pots or cases, consists of the native earths called peat or bog earth, and sandy loam, mixed in about equal proportions, with a further admixture equal to an eighth of the whole mass for the coarser sorts, and of a fourth of the whole mass for the more delicate sorts, of any pure granulated silicious matter, which is used for the purpose of preventing the too close adhesion and consolidation of the particles; the clean white sand called Reigate sand is that most generally employed. They are not benefited by manure.

The supply of water to Ferns under artificial conditions is a very essential matter; they must never lack moisture, or their fragile texture shrinks as before a burning blast; nor, with few exceptions, must the soil about them be kept continually wet with stagnant water; indeed, stagnant water is in all cases much better avoided.

The head-quarters of Ferns are the humid forests of tropical islands, in some of which they acquire a giant size, and in their tree-like habit become rivals to the noble Palms. The tree Ferns are not, however, very numerous compared with those of dwarfer growth.

The dwarf herbaceous Ferns are characteristic of the temperate and colder zones; but even in the temperate regions some of these herbaceous Ferns attain considerable height, as is the case with the common Bracken, which, in the hedgerows of sheltered rural lanes in the south of England, reaches the height of eight or ten feet, and assumes the most graceful habit that can be conceived.

Wherever the Ferns occur, whether it be the herbaceous species of temperate climates, or the arborescent species of the equatorial regions, or the epiphytal species which clothe the trunks and branches of the trees in tropical forests, they add a marked and peculiar character of beauty and luxuriance to the scenery, and that to an extent which is not realized by any other race of plants.

The uses of Ferns do not form a long catalogue. Two of our common native species, the Filix-mas and the Bracken, especially the former, have the reputation of being remedies against intestinal worms; their properties being bitter and astringent. Another native Fern, the Royal Fern, is much used as a rustic vulnerary and as an application to sprains or bruises. The Maidenhair is employed in the form of *Capillaire*, which is prepared by pouring boiling syrup over the fronds, and flavouring the infusion with orange flowers. This preparation is considered pectoral, though, if too strong, it is said to be emetic. The common Adder's-tongue is gathered by country-people for the preparation of adder's-spear ointment, which is a popular remedy for recent wounds.

Both the common Bracken and the Male Fern abound in alkali, which is applied to various economic uses, as the manufacture of soap and glass. From their astringency they are employed sometimes in the dressing of leather, &c. These species have also been used in the preparation of beer.

Ferns are amongst the best of all plants for preservation, in the dried state, in the form of an herbarium; for, in addition to their elegant appearance when nicely arranged on sheets of white paper, they are less liable than most plants to the attacks of the destructive pests in the shape of insects, which commit great havoc among dried plants in general.

The plants should be dried quickly, under moderately heavy pressure, among sheets of absorbent paper, which must be replaced daily by dried sheets as long as the plants continue to give out moisture. The thicker the bulk of paper placed between the specimens whilst under pressure, the better. Two or three changes will generally be sufficient, if the substituted sheets be in each case perfectly dry.

The smaller growing kinds should be gathered, if possible, in the tufts as they grow, preserving the whole mass of fronds, with the stem and roots, the fronds being spread out in an easy and graceful form, and as far as possible kept quite flat. If entire tufts cannot be obtained, and single fronds have to be substituted, they should be taken quite to the base, and must be removed from the stem with care, so that the scales, or hairs, or farinose powder, which may be present on the stalk, may be preserved equally with the frond itself.

Of larger-growing species, single fronds only are manageable, and these, when of larger size than the folios in which the specimens are to be kept, must be folded to somewhat less than the length of the papers, *whilst yet fresh*. Of the gigantic species, portions only of the fronds, corresponding in size with the paper to be used, can be preserved; but all our native species, except in cases of extreme luxuriance, may, we believe, with a little judgment in the selection of specimens, be folded so as to allow of their being preserved in ordinary folios measuring eighteen inches by twelve inches, or thereabouts.

It is sometimes recommended to select specimens with the fructifications mature. We should rather, as a general rule, advise their being gathered just before the masses of spores reach their full growth. If, however, more than a single specimen of each kind is preserved, the perfectly mature and the incipient states of fructification should also be gathered; but in the majority of cases the intermediate state will afford the best materials for subsequent examination and recognition. Certainly the fructification is to be preferred in an early rather than a late stage of development.

Of course, when the species produces two or more kinds of fronds, examples of each must be preserved, as, for instance, in the *Allosorus crispus*, the fertile fronds of which alone would convey but a very indifferent notion of the plant.

After being thoroughly dried under pressure, the specimens, according to their size, should be arranged, singly if large, or in groups resembling the natural tufts if sufficiently small, on one side only of a series of sheets (technically half-sheets, *i.e.*, single leaves) of stout white paper, to which they should be fastened by a few thread ties, or gummed straps, or they may be fastened down with glue. The specimens, no doubt, admit of a much more convenient and searching examination when kept loose in a folded sheet of paper; but if there should be frequent occasion to handle such loose specimens, they will be found much more liable to become injured and broken than such as are fastened to the paper.

The specimens should be fully labelled, the labels giving at least their names, the locality where gathered, and the date; and the labels should be fixed so as to be readily referred to by turning up one of the corners of the sheets of paper.

The papers to which the specimens are affixed are to be enclosed in paper covers, formed of whole sheets, *i.e.*, two leaves, each genus being put in a separate cover. These covers should be placed either on the shelves of a cabinet, or in drawers, or in any convenient place where they may be protected from dust, and preserved against the attacks of insects and other casualties

CHAPTER III.

TABULAR ARRANGEMENT OF THE BRITISH FERNS—GENERA—SPECIES—VARIETIES.

THE Ferns, it will be recollected, are flowerless plants, bearing seed-vessels (spore-cases) on their leaves (fronds), and these spore-cases are placed either on the back of the frond (dorsal) or on their margins (marginal). The British Ferns belong to groups which are called *Polypodiaceæ*, *Osmundaceæ*, and *Ophioglossaceæ*.

I. POLYPODIACEÆ. This group consists of Ferns having the leaves rolled up in a circinate or crosier-like manner while young; and the spore-cases girt with an elastic ring, and bursting by an irregular transverse cleft. It comprises several lesser groups or sections called *Polypodieæ*, *Gymnogrammeæ*, *Aspidieæ*, *Asplenieæ*, *Blechneæ*, *Pterideæ*, *Adianteæ*, *Cystopterideæ*, *Woodsieæ*, and *Hymenophylleæ*.

§ POLYPODIEÆ=*Ferns whose round clusters of spore-cases have no special membranous cover (indusium).*

1. *Polypodium*=Dorsal-fruited Ferns, with the circular sori exposed, *i.e.* without covers.
2. *Allosorus*=Dorsal-fruited Ferns, with the roundish sori becoming laterally confluent beneath the reflexed, unaltered margins of the frond.

§ GYMNOGRAMMEÆ=*Ferns whose linear clusters of spore-cases have no special cover.*

3. *Gymnogramma*=Dorsal-fruited Ferns, having the sori linear, forked, naked.

§ ASPIDIEÆ=*Ferns whose sori have special indusia, of a circular or roundish form, and springing here and there, from the back of the veins.*

4. *Polystichum*=Dorsal-fruited Ferns, having circular peltate indusia, attached by their centre.
5. *Lastrea*=Dorsal-fruited Ferns, having reniform indusia, attached by their indented side.

§ ASPLENIEÆ=*Ferns whose sori have special indusia, of an oblong or elongated form, and springing from the sides of the veins.*

6. *Athyrium*=Dorsal-fruited Ferns, having oblong reniform

indusia, attached by their concave side, the detached side fringed with hair-like segments.

7. *Asplenium*=Dorsal-fruited Ferns, having the indusia straight and elongate, and attached by the side towards the margin of the pinnæ or pinnules.

8. *Scolopendrium*=Dorsal-fruited Ferns, having the sori elongate, and proximate in parallel pairs, the indusia opening along the centre of the twin sori.

9. *Ceterach* = Dorsal-fruited Ferns, having the indusia obsolete, and the sori hidden among densely imbricated, rust-coloured, chaffy scales.

§ BLECHNEÆ=*Ferns whose sori have special indusia, forming longitudinal lines between the midrib and margins of the leaflets or divisions of the frond.*

10. *Blechnum*=Dorsal-fruited Ferns, having the spore-cases in a continuous line between the midrib and margin of the divisions of the frond, covered by linear indusia.

§ PTERIDEÆ = *Ferns, the margin of whose frond is soriferous, and continuously or interruptedly changed into a special indusium.*

11. *Pteris*=Dorsal-fruited Ferns, having the spore-cases in a continuous line at the edge of the frond, beneath indusia, formed of the altered margin.

§ ADIANTEÆ=*Ferns, the margin of whose fronds bears reflexed lobes, which are changed to indusia, and bear the spore cases on their under surface.*

12. *Adiantum*=Dorsal-fruited Ferns, having the spore-cases in patches, on the reflexed, altered apices of the lobes of the fronds, which form indusia.

§ CYSTOPTERIDEÆ=*Ferns whose sori have special ovate indusia affixed behind the sori, and inflected hood-like over them.*

13. *Cystopteris* = Dorsal-fruited Ferns, having cucullate or hooded semi-involucriform indusia, attached by their broad base.

§ WOODSIEÆ=*Ferns whose sori have special involucriform or semi-involucriform indusia, roundish, and springing from the back of the veins.*

14. *Woodsia*=Dorsal-fruited Ferns, having the indusia involucriform, *i.e.*, attached beneath the sori, and divided at the margin into hair-like incurved segments.

§ HYMENOPHYLLEÆ = *Ferns whose sori are produced around the ends of veins projecting from the margin, and surrounded by urn-shaped or two-valved membranes.*

15. *Trichomanes*=Marginal-fruited Ferns, having the sori surrounded by urn-shaped expansions of the frond.
16. *Hymenophyllum*=Marginal fruited Ferns, having the sori surrounded by two-valved expansions of the frond.

II. OSMUNDACEÆ. This group consists of Ferns having the young leaves circinate, the spore-cases destitute of an elastic ring, and bursting vertically by two regular valves.

17. *Osmunda*=Marginal-fruited Ferns, having the regular valved spore-cases in irregular, dense, branching clusters, terminating the fronds.

III. OPHIOGLOSSACEÆ. This group is composed of Ferns having the young leaves folded up straight, the spore-cases destitute of an elastic ring, and two-valved.

18. *Botrychium*=Marginal-fruited Ferns, having the spore-cases in irregularly branched clusters, on a separate branch of the frond.
19. *Ophioglossum*=Marginal-fruited Ferns, having the spore-cases sessile in two-ranked simple spikes terminating a separate branch of the frond.

Genus 1. POLYPODIUM, *Linnæus.*

1. *P. vulgare.*—Fronds oblong pinnatifid.

    Var. semilacerum.—Fronds pinnatifid, lower segments again pinnatifid, upper fertile.

    Var. cambricum.—Fronds pinnatifid, segments again pinnatifid, all barren.

2. *P. Phegopteris.*—Fronds pinnate below; pinnæ pinnatifid.
3. *P. Dryopteris.*—Fronds ternate, deltoid, glabrous.
4. *P. Robertianum.*—Fronds subternate, elongate, deltoid, glandular-mealy.
5. *P. alpestre.*—Fronds bipinnate, lanceolate.

    Var. flexile.—Fronds narrow, flaccid; pinna deflexed.

Genus 2. ALLOSORUS, *Bernhardi.*

1. *A. crispus.*—The only British species.

Genus 3. GYMNOGRAMMA, *Desvaux.*

1. *G. leptophylla.*—The only British species.

Genus 4. POLYSTICHUM, *Roth.*

1. *P. Lonchitis.* Fronds pinnate, spiny-serrate.
2. *P. aculeatum.*—Frond bipinnate, rigid; pinnules ovate subfalcate, auricled, acute, acutely wedge-shaped at the base, nearly all distinct.
   Var. lobatum.—Fronds narrower, pinnules nearly all decurrent.
3. *P. angulare.*—Fronds bipinnate; lax; pinnules oblong or ovate subfalcate, auricled, bluntish or acute, obtusely angled at the base, stalked.
   Var. subtripinnatum.—Fronds ample; lower pinnules deeply pinnatifid, otherwise normal.
   Var. proliferum.—Fronds bearing bulbils; pinnules narrow, acute, deeply pinnatifid, with distant attenuate lobes.
   Var. imbricatum.—Fronds narrow; pinnules oblong-obtuse, overlapping.
   Var. alatum.—Fronds normal in outline; pinnules connected by a broad wing, which obliterates the stalk.

Genus 5. LASTREA, *Presl.*

1. *L. Thelypteris.*—Fronds pinnate, not glandular; sori submarginal on more or less contracted fronds; caudex creeping.
2. *L. montana.*—Fronds pinnate, covered with sessile glands; caudex tufted.
3. *L. Filix-mas.*—Fronds sub-bipinnate or bipinnate, broadly lanceolate; indusium plain.
   Var. incisa.—Larger, pinnules elongate, with deep serrated incisions.
   Var. paleacea.—Larger, pinnules truncately-obtuse, entire; stipes very scaly.
   Var. pumila.—Smaller, pinnules contracted, confluent.
4. *L. rigida.*—Fronds bipinnate, without spinulose serratures, glandular; indusium fringed with glands.
5. *L. cristata.*—Fronds narrow linear, pinnate or sub-bipinnate, pinnules oblong, with aristate teeth, the posterior and anterior ones nearly equal; scales ovate, pale.
   Var. uliginosa.—Fronds (fertile) narrow linear-lanceolate, bipinnate at the base, pinnules oblong, acute, with aristate teeth, the posterior and anterior ones nearly equal; sterile fronds as in 5; scales ovate, pale.

Var. spinulosa.—Fronds oblong-lanceolate, bipinnate, with spinulose serratures; posterior pinnules much largest; scales ovate, pale.

6. *L. dilatata.*—Fronds oblong, or ovate-lanceolate, bi-tri-pinnate, with spinulose serratures; scales lanceolate.

Var. dumetorum.—Fronds oblong-ovate or ovate-triangular, dwarf, very glandular; scales pale, two-coloured, narrower.

Var. collina.—Pinnules ovate, blunt, bluntly mucronate-serrate; scales dark-centred.

Var. Chanteriæ.—Fronds lanceolate, narrowed below, and caudate at the apex; pinnæ and pinnules distant.

Var. glandulosa.—Fronds lanceolate-ovate, tall, very glandular; scales pale two-coloured, broader.

7. *L. æmula.*—Fronds triangular, bipinnate, pinnules concave above; scales lanceolate, laciniated.

Genus 6. Athyrium, *Roth.*

1. *A. Filix-fœmina.*—The only British species. Fronds bi-pinnate; pinnules flat, linear-oblong, lobed.

Var. rhæticum.—Pinnules narrow, distinct, linear, convex, lobed.

Var. latifolium.—Pinnules broad ovate, crowded, irregularly lobed.

Var. marinum.—Fronds narrowed to the base, decumbent, pinnules oblong, with simple marginal teeth, rachis winged.

Genus 7. Asplenium, *Linnæus.*

1. *A. septentrionale.*—Frond linear-lanceolate, two-three-cleft.
2. *A. germanicum.*—Fronds linear, alternately pinnate, pinnæ narrow wedge-shaped; indusium entire.
3. *A. Ruta-muraria.*—Fronds bipinnate, pinnules wedge-shaped at the base; indusium jagged.
4. *A. viride.*—Fronds linear, pinnate, rachis green above.
5. *A. Trichomanes.*—Fronds linear, pinnate, rachis black throughout.

   Var. incisum.—Pinnæ deeply lobed.
6. *A. marinum.*—Fronds pinnate, rachis winged.
7. *A. fontanum.*—Fronds bipinnate, narrow lanceolate, rachis winged, smooth.
8. *A. lanceolatum.*—Fronds bipinnate, broad lanceolate, rachis not winged, scaly.

Var. microdon.—Pinnæ only lobed or subpinnate at the base, wavy.

9. *A. Adiantum-nigrum.*—Frond bipinnate, triangular; segments wedge-oblong.

Var. acutum.—Fronds tripinnate, triangular, much acuminate; segments linear, acute.

Genus 8. CETERACH, *Willdenow.*

1. *C. officinarum.*—The only British species.

Genus 9. SCOLOPENDRIUM, *Smith.*

1. *S. vulgare.*—The only British species. Fronds strap-shaped, entire.

Var. polyschides.—Fronds long, narrow, irregularly lobed, fertile.

Var. marginatum.—Fronds long, narrow, lobed, with a double, *i.e.* split margin, fertile.

Var. crispum.—Fronds elongated, much undulated at the margin, usually barren.

Genus 10. BLECHNUM, *Linnæus.*

1. *B. Spicant.*—The only British species.

Genus 11. PTERIS, *Linnæus.*

1. *P. aquilina*—The only British species.

Genus 12. ADIANTUM, *Linnæus.*

1. *A. Capillus-Veneris.*—The only British species.

Genus 13. CYSTOPTERIS, *Bernhardi.*

1. *C. fragilis.*—Fronds lanceolate, bipinnate, pinnules ovate, acute, toothed or lobed; sori central.

Var. dentata.—Pinnules ovate, obtuse, bluntly toothed, distinct; sori marginal.

Var. Dickieana.—Pinnules broad, obtuse, slightly blunt-toothed, overlapping; sori marginal.

2. *C. regia.*—Fronds lanceolate, subtripinnate, segments linear.

3. *C. montana.*—Fronds triangular, tripinnate.

Genus 14. WOODSIA, *R. Brown.*

1. *W. ilvensis.*—Fronds lanceolate, hair-scaly; pinnæ oblong, deeply lobed.

2. *W. alpina.*—Fronds linear, almost smooth; pinnæ bluntly triangular, lobed.

Genus 15. TRICHOMANES, *Linnæus.*

1. *T. radicans.*—The only British species.

Genus 16. HYMENOPHYLLUM, *Smith.*

1. *H. tunbridgense.*—Pinnæ vertical; involucres compressed, serrate.
2. *H. unilaterale.*—Pinnæ deflexed; involucres inflated, entire.

Genus 17. OSMUNDA, *Linnæus.*

1. *O. regalis.*—The only British species.

Genus 18. BOTRYCHIUM, *Linnæus.*

1. *B. Lunaria.*—The only British species.

Genus 19. OPHIOGLOSSUM, *Linnæus.*

1. *O. vulgatum.*—Fronds ovate.
2. *O. lusitanicum.*—Fronds very small, linear-lanceolate, obtuse, fleshy.

CHAPTER IV.

THE POLYPODIES.

THE Polypodies are named POLYPODIUM by botanists. They are known from all the other British Ferns, by their having the spore-cases arranged in little round patches placed on the back of the frond, these patches not being at any stage of their development covered by the membranous film called an indusium; hence they are said to be naked, or non-indusiate. This family includes five distinct kinds, and there are many variations of the common sort. In some of the species the fronds continue green through the winter, so that they are evergreen; while in others they last but from spring to autumn in each succeeding year.

The technical name *Polypodium* given to this family is derived from the Greek, and literally means *many-footed.* This has been explained to apply to the branching of their creeping stems, and also to the protuberances on them which, in the earlier stages of development, have some supposed resemblance to the feelers of Polypes.

COMMON POLYPODY.—[*Plate* I.]

The name which this plant bears among botanists is *Polypodium vulgare.* It has also been called *Ctenopteris vulgaris.* The Common Polypody grows with a creeping stem nearly as thick as one's finger, and covered over with pale brown chaffy taper-pointed scales. From the upper side of this stem, or rhizome, spring the fronds, and from its lower side chiefly the branching fibrous roots by which it clings to its support. The fronds, if exposed to frost, perish; but if at all sheltered, they remain green during winter, and until after young ones have been produced, which happens generally towards the end of May. The stipes, or stalk, of the full-grown fronds is usually nearly equal in length to the leafy portion; the entire frond measuring from six to eighteen inches in length. The leafy part of the frond is lance-shaped in outline, but cut in from the margin along both sides nearly as far as the midrib or rachis, and thus becomes what is called pinnatifid. The portions into which it is divided are called the lobes, or segments, or divisions of the frond; and in this case they are usually oblong in form, generally rounded at the end, but sometimes tapering to a blunt point, and occasionally notched along the margin. Each lobe has a slightly wavy midvein, producing alternate lateral veins (venules), which generally have about four veinlets or little veins disposed alternately; it is the lowest of these veinlets, on the side towards the point of the lobe, which produces the sorus when it is present; the rest, which are barren, terminate in club-shaped heads, which are very readily seen when a fresh frond is held between the eye and a strong light. Most of the fronds of this kind of Fern produce fructification, which, however, is usually confined to the upper half of the fronds, and has generally become mature by the end of September.

It is an evergreen Fern, growing abundantly on pollard trunks, mossy banks, moist rocks and walls, and old thatched roofs; and is pretty generally distributed over the United Kingdom. It may be recognised by the comparatively large circular patches of golden spore-cases; no other native sort having the fructification at all similar in appearance.

The most remarkable variety is *cambricum*, commonly called the Welsh Polypody, named *Polypodium cambricum*, by Linnæus. In this the lobes of the frond are broader, and, instead of being simple, are deeply and irregularly lobed a second time, the segments being rather sharply toothed. This form, which is certainly only a variety of the common Polypody, is always found without fructification. Under slight shelter, where its fronds are persistent, it is one of the most beautiful of what are called hardy Ferns. The Irish Polypody, an equally elegant form, called *semilacerum*, is found in Ireland and elsewhere; in this the lower half of the fronds are a second time lobed, and the upper half usually fertile, and not twice-lobed. There is a form resembling *cambricum*, but differing in yielding fertile fronds, which is called *omnilacerum*.

The plant varies in many ways. In some forms the lobes are more or less regularly two-cleft at the apex; in others the margins are deeply saw-edged, or wavy or irregular; in some the margin has rounded notches; and the segments are occasionally drawn out to a long narrow point. A complete enumeration of the varieties of this and the other of our native species of Ferns, which would occupy too much space here, will be found in the new edition of our *Handbook of British Ferns.*

The species and its varieties grow readily under cultivation, either planted in pots, or on rockwork in a shady situation. They should have a light open soil.

MOUNTAIN POLYPODY, OR BEECH-FERN.

FIG. 1.

Polypodium Phegopteris.

The proper name of this plant is *Polypodium Phegopteris.* It has, however, been variously called *Lastrea Phegopteris, Polystichum Phegopteris, Gymnocarpium Phegopteris, Phegopteris polypodioides,* and *Phegopteris vulgaris.*

The plant has a slender but extensively creeping and slightly scaly stem, producing black fibrous roots. About May these stems throw up delicate hairy pale green fronds, which, when full grown, measure from six inches to a foot in height. The stipes, which is fleshy and very brittle, is generally twice as long as the leafy part of the frond. Near its base are a few small almost colourless scales. The fronds are triangular, extended into a long narrow point. In the lower part they are pinnate; but this distinction of the parts is seldom carried beyond the two lowest pairs of branches, those of the upper portions of the frond being connected at the base, in what is technically called a pinnatifid manner:

hence this Fern is said to be subpinnate, which, in this case, means partially pinnate, or pinnate at the very base only. The pinnæ have a narrow and acutely lance-shaped outline, and are deeply pinnatifid; they usually stand opposite each other in pairs, the lowest pair being directed downwards, towards the root, and set on at a short distance from the rest. The united bases of the pairs of the other pinnæ, when they happen to stand exactly opposite each other, exhibit a cross-like figure more or less obvious; and by this mark, in conjunction with the triangular outline and subpinnate mode of division, this species may be known from the other British Polypodies. The veins in the lobes of the pinnæ are pinnate; that is to say, there is a slender midvein, from which alternate venules, mostly unbranched, extend to the margin, those near the base of the lobes bearing each one small circular sorus near their extremity—the fructification thus becoming almost marginal.

This is a somewhat fragile plant, enduring no longer than till autumn, or the appearance of the first frosts. It grows wild in moist mountainous situations and in damp woods, often common enough where present, but rather limited in its range; occurring however in England to the southward, westward, and northward; pretty generally distributed in Scotland; but rarely met with in Ireland.

It is a very delicate and graceful Fern for pot-culture or for a Wardian case, and requires plenty of percolating moisture. On the damp, shady sides of sheltered artificial rock-work, in the open air, it grows with tolerable vigour.

SMOOTH THREE-BRANCHED POLYPODY, OR OAK-FERN.

This species is named *Polypodium Dryopteris* by botanists. It is sometimes, but less frequently, called *Lastrea Dryopteris*, *Polystichum Dryopteris*, *Gymnocarpium Dryopteris*, and *Phegopteris Dryopteris*.

The fronds of this delicate little Fern grow from a slender creeping stem, which often forms densely matted tufts. They are quite smooth, and of a bright light green colour, supported by stipes, which are usually about twice as long as the leafy part, and are slender, brittle, and dark-coloured. The outline is almost pentagonal, the frond being divided into three branches, each of which is of a triangular form. One peculiarity about this species, which is in a slight degree shared by its near ally *P. Robertianum*, is the deflexion of the rachis at the point where the lateral branches of the frond take their rise, but this feature is much more obvious in *P. Dryopteris.* The fronds are divided so that each branch is pinnate at the base, and pinnatifid towards its point; the pinnæ are also pinnate at their base, then pinnatifid, becoming acute and nearly entire at the point; the pinnules and ultimate lobes are oblong and obtuse. The pair of pinnules at the base of each pinna, close to the principal rachis, are placed so that when the pinnæ are exactly opposite they stand in the form of a cross; the two towards the apex of

FIG. 2.

Polypodium Dryopteris.

the branch being smaller than the opposite pair, and more nearly parallel with the rachis. The pinnules, or lobes, have a rather wavy midvein, from which the venules branch out alternately, being, in those of moderate size, simple, with a sorus near their extremity, and in those which are larger and more compound, branched, with a sorus on the lower branch. The fructification is very unequally produced in different seasons and localities, being sometimes crowded, and at other times very sparingly scattered over the fronds.

This is at once known among the Polypodies by having

its fronds smooth and divided into three branches. When the fronds are but partially developed this latter characteristic is very obvious; for the three branches are seen to be rolled up separately at the end of these little wire-like stalks, the whole supported by one which is longer and stouter. It is a slender and delicate plant, its height being commonly not more than six inches, often less, though sometimes more, and its texture fragile. Hence it is at once destroyed by frost, and soon becomes rusty and withered by exposure to heat and drought. When growing in a cool, shady situation, however, it continues fresh and cheerful-looking from April, when it usually starts into growth onwards until it is affected by autumnal cold. In pots, in Wardian cases, or on sheltered shady rock-work, it is alike desirable for cultivation.

P. Dryopteris is not an uncommon species, but it occurs only in mountainous situations and the drier parts of damp woods: in England mostly in the north; in Scotland distributed pretty generally; very rare in Ireland.

LIMESTONE POLYPODY.

The proper name of this Fern is *Polypodium Robertianum*, but it is more commonly known as *P. calcareum*, that name having been much used by English writers. Other names applied to it are *Lastrea Robertiana*, *Phegopteris calcarea*, and *Gymnocarpium Robertianum*.

The *Limestone Polypody* grows from six inches to a foot in height. The fronts are nearly triangular, with the base shorter than the sides, the stipes about equalling the leafy portion in length. They are partially three-branched, but the lateral branches are much smaller than the central one, and attached to the stipes by a more slender rachis. The lower branches are pinnate, with pinnatifid pinnæ; the upper branch pinnate, with its lower pinnæ again pinnate, and the upper ones pinnatifid, as also is the apex of the frond and of the lower branches. The pinnules, or lobes, have a distinct midvein, with simple or slightly branched venules, near the termination of which, in a marginal series, the sori are produced.

This Fern is known from *P. Dryopteris*, to which it is so nearly related, that some botanists do not consider it distinct, by having its fronds less decidedly, though somewhat three-branched, and by having its surface covered with small stalked glands, which give a mealy appearance to every part. In addition to these points of difference, the fronds of this are of a dull deep green, more rigid, and without the marked deflection of the rachis so obvious in *P. Dryopteris*; and the young fronds, instead of being rolled up in three little balls, have their pinnæ all rolled up separately. The glandular surface of the frond is very readily seen with a good pocket-lens, which is a necessary aid to the study of Ferns.

FIG. 3.

Polypodium Robertianum.

This is one of the few Ferns which are found in calcareous or chalky soils. It is rare, and local in its distribution, being, we believe, almost confined to rocky limestone districts, and occurring chiefly in the northern and western parts of the island.

In cultivation, this species does not require so much moisture and shade as most other Ferns, but a limestone soil is not at all essential to its well-being.

ALPINE POLYPODY.

This plant is the *Polypodium alpestre* of botanists; it has also been called variously, *Aspidium alpestre*, *Athyrium alpestre*, *Pseudathyrium alpestre*, and *Phegopteris alpestris*.

FIG. 4.

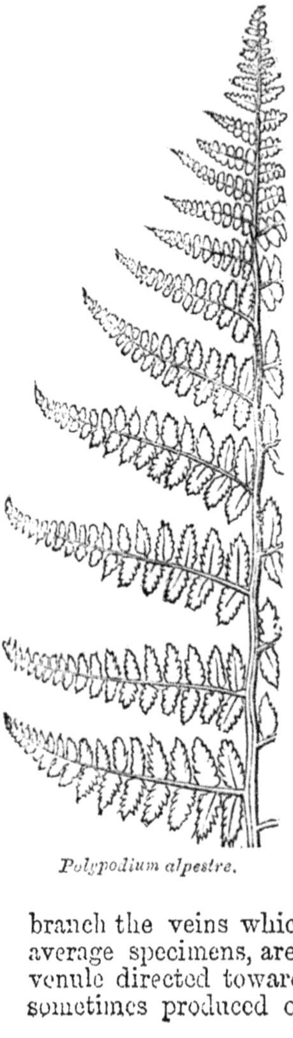

Polypodium alpestre.

FIG. 5.

Polypodium alpestre: pinnule magnified.

The *Alpine Polypody* has a short decumbent root-stock, producing fronds in tufts from the crown. They are from six inches to three feet or more in height, broadly lance-shaped and attached by comparatively short stipes, clothed with broadish-pointed membranous scales. They are bipinnate, or sometimes subtripinnate. The lower pinnæ are gradually shorter, so that the outline is truly lanceolate. The pinnæ are linear-lanceolate, taper pointed, spreading at an obtuse angle with the rachis. The pinnules are numerous, ovate-oblong, acute, variously pinnatifid, the segments notched with sharp, coarse teeth; rarely the pinnules are ovate-lanceolate, and in the most vigorous fronds they are so deeply pinnatifid as to become almost tripinnate. The pinnules have a slightly wavy midvein, from which alternately branch the veins which ramify in the lobes; these veins, in average specimens, are pinnately branched, with a simple venule directed towards each marginal tooth. The sori are sometimes produced only on the lowest anterior venule of each lobe, and they then form a series on each side the mid-vein; but sometimes more of the venules are fertile, and the sori then range in short lines near the margin of the lobes.

This Fern has so remarkably the aspect of the common Lady Fern, that although common on the Scotch mountains, it has been till very recently overlooked, the plants having been supposed to belong to that species. It would appear to be plentiful on the higher parts of the mountainous districts of the counties of Perth, Forfar, and Aberdeen, in company with the Lady Fern in its lower range, but without it at higher elevations. The fronds appear in May, and perish early in autumn.

The very distinct and remarkable variety called *flexile*—called by some *Pseudathyrium flexile*, differs in its lax spreading habit, narrow fronds, short deflexed pinnæ, and fewer pinnules. The plant as seen in cultivation produces stalkless or almost stalkless fronds, which often bear their sori abundantly at the base, but scarcely if at all upwards. These latter marks—the sessile fronds, and basal sori—disappear in some instances both in the wild and cultivated specimens. A very imperfect indusium has been observed on some of the sori in the plant under culture, but occurs on the least perfect sori. The sori, as also sometimes happens in *P. alpestre* itself, is not in all cases punctiform, but occasionally, though rarely, lateral on the veins. It was first found in 1852, in Glen Prosen, by Mr. Backhouse and Mr. Westcombe; and the same botanists again found it plentifully in the same district of the Clova Mountains, in the summer of 1855.

CHAPTER V.

THE ROCK BRAKES.

THE technical name of the family to which the Rock Brakes belongs, is *Allosorus*. The family is known from all other British Ferns by the coincidence of the following features:—It bears fronds of two kinds, one being leafy and barren, *i.e.*, without sori; the other contracted, and bearing sori, and hence called fertile; then the edges of the lobes of the fertile fronds are rolled under (which is what gives them the contracted or narrowed appearance), and cover the sori in the stead of an indusium; and moreover, the sori when young form distinct circular clusters beneath this recurved margin of the frond; but as they grow they join laterally (in technical language, become confluent), forming two lines of fructification lengthwise the segments of the fronds.

The name *Allosorus* is compounded from the Greek, and comes from two words, which means *various* and *a heap;* the intention no doubt being to indicate the variation in the arrangement of the sori, occurring among the plants originally thought to belong to this family. It may also apply to the apparent difference of arrangement in the sori of this plant at different stages of development, the young sori forming distinct roundish patches, and the older becoming effused into larger shapeless masses.

ROCK BRAKES, OR MOUNTAIN PARSLEY. [*Plate* II. fig. 1.]

The scientific name *Allosorus crispus* is that preferred by us for this Fern. Others prefer to use one of the following—*Cryptogramma crispa, Pteris crispa, Osmunda crispa,* the first of these three being however now mostly used, when *Allosorus* is passed over.

This elegant little plant, which has considerable first-sight resemblance to a tuft of parsley, and is hence sometimes called Mountain Parsley, grows in a dense tuft, throwing up its fronds in May or June, and losing them in the course of the autumn. The fronds average about six inches in height, and are generally somewhat three-cornered in outline, with a longish, slender, smooth stalk. They are, as already stated, of two kinds; both kinds being twice or thrice pinnate, and of a pale green colour. The segments into which the sterile fronds are cut, are more or less wedge-shaped, and notched or cleft at the end. The fertile fronds have the segments of an oval or oblong or linear form. The divisions of the fertile frond have a slightly tortuous midvein, producing simple or forked venules, which extend nearly to the margin, each, for the most part, bearing near its extremity a circular sorus. There is no true indusium, but the sori are covered by the reflexed and partially bleached margins, which sometimes almost meet behind, so that the spore-cases are quite concealed. The patches of spore-cases are at first distinct, but ultimately they spread out and become more or less confused and blended.

The Rock Brakes is a mountain Fern, choosing to grow in stony situations. It is comparatively rare and local; most abundant in the north of England and Wales, and less plentiful in Scotland and Ireland.

This plant grows readily in pots, and also n a Wardian case not too confined; for either of these modes of cultivation its small size and elegant aspect render it a very desirable object. It is, however, very impatient of root moisture.

CHAPTER VI.

THE GYMNOGRAM.

THIS plant belongs to a family of which nearly all the species are tropical. One small annual kind has, however, been found to inhabit Jersey, and is thus brought—politically, not geographically—within the limits of the British Flora. The peculiarity in this genus is that the spore-cases, which are not covered, are scattered in lines along the veins, extending, in many cases, below the point where the latter separate into branches. The sori thus become what is technically called linear and forked, as well as naked.

The name is derived from two Greek words, signifying *naked* and *a line;* the lines of spore-cases, without covering membranes, which are the peculiar features of the group, are thus distinctly indicated.

SLENDER GYMNOGRAM. [*Plate* II. fig. 2.]

This little Fern bears the scientific name of *Gymnogramma leptophylla*, which is the one most commonly used. It has, however, received several others, such as *Polypodium leptophyllum*, *Acrostichum leptophyllum*, *Grammitis leptophylla*, *Asplenium leptophyllum*, &c. The apparent contradiction arising from the use of so many conflicting names, is explained by the different stages of growth in which the plant has been examined by the botanists who imposed them, as well as by the less precise information existing at the date when many of them were given.

The *Gymnogramma leptophylla* is a small Fern, of short duration, springing up from the spores in the autumn of each year, attaining maturity early in the following summer, and becoming afterwards quickly dried up and disappearing. Each plant consists of a tuft of about half a dozen fronds, of which the latest, and largest, are from three to six inches high, and bear fructification. Some of the early fronds are short and fan-shaped, divided only into two or three lobes; the succeeding ones grow an inch or two long, and become pinnate with obliquely fan-shaped three-lobed pinnæ; and finally, the fertile fronds which are taller and more erect in growth, are ovate, and two or three times pinnate; the pinnæ being alternate, ovate, with alternate pinnules; the ultimate pinnules roundish, wedge-shaped, three-lobed at the apex, the lobes rather distinct, and usually notched at the end. The veins in each pinnule are branched, so that one of the small veins proceeds towards each of the teeth into which the pinnule is divided; and the spore-cases are borne along these branches of the veins. The lines of sori on the pinnules often become united into a mass, after they have been some time developed.

Though a minute species, this fern is widely scattered over the face of the globe; it is plentiful in many parts of the South of Europe, and extends as far northwards as Jersey.

It grows readily, as an annual, sown in sandy loam, and kept in a rather warm damp situation.

CHAPTER VII.

THE SHIELD FERNS.

The Shield Ferns are called *Polystichum*. They form a small and very distinct group of evergreen Ferns, some of which rank among the most beautiful of our native species. They once formed part of the genus *Aspidium*, in consequence of their having round seed-patches covered by a scale; but that family, as far as the British kinds are concerned, is broken up by modern botanists, in whose ideas we coincide, into two groups, called *Polystichum* and *Lastrea*. The *Polystichums* are known from the allied *Lastreas*, by their having the scale-like cover of the sori circular, without a lateral notch, its attachment being by a little stalk in the centre of the under side. This form of attachment is technically called peltate. To a practised eye they are also known by their more rigid texture, and by their having altogether a more spiny appearance than even the spinulose species of *Lastrea*. Our alpine species, *P. Lonchitis*, is strictly evergreen, and the other species acquire this character when in a sheltered situation; but if they are much exposed, the fronds are killed by severe frosts. In general, however, they all retain their fronds without much disfigurement from frost, quite through the autumn, and often far into winter. The British species of *Polystichum* are three in number, but of one of them there are very many highly interesting varieties; for an account of which we must again refer to our *Handbook of British Ferns.*

The name *Polystichum* is compounded of two Greek words, signifying *many*, and *order;* and it is applied to these plants in allusion to the numerous regular lines of sori, which are seen distributed over the fronds.

ALPINE SHIELD FERN, OR HOLLY FERN.

This Fern bears the scientific name *Polystichum Lonchitis.* Synonymous with this, but less often used, are *Polypodium Lonchitis*, and *Aspidium Lonchitis.*

FIG. 6.

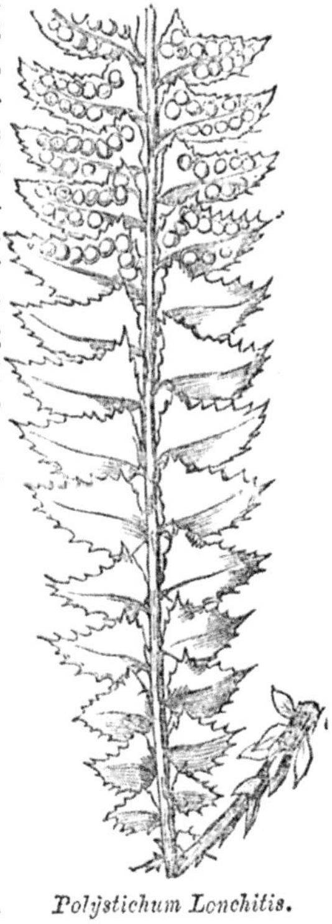

Polystichum Lonchitis.

This Holly Fern is a rigid and prickly-looking species, whence comes one of its English names. It has a scaly tufted stem, from the crown terminating which, the young fronds are produced early in each spring. The fronds remain fresh and vigorous, until after those of the succeeding year are developed, so that the species is truly evergreen in its habit of growth. The size of the fronds is very variable; sometimes they are not more than six inches long, and cultivated plants do not often much exceed this stature. In damp and but slightly elevated situations it becomes more luxuriant, the fronds sometimes attaining a foot and a half in length, and then having a vigour and robustness of aspect never acquired, as far as we know, in cultivation, at least in England. The climate of Ireland seems more congenial to it. The fronds are narrow in outline, their figure being linear-lanceolate; they are once pinnate, the pinnæ being short, crowded, and between sickle-shaped and crescent-shaped, the upper side at the base having an ear-shaped projection, called an auricle, the lower side being, as it were, cut away. The margin is set with spinous teeth. The veins are twice branched, the branches extending to the margin without joining with others. The clusters of spore-cases form a line parallel with and on each side of the midrib, and are covered each by a membranous circular scale, which is attached by a short central stalk.

This is a true rock-Fern, occurring on the bleak mountains of Scotland and in the milder climate of Ireland, as well as, rarely, in the north of England and Wales.

The Holly Fern is very distinct, and, when vigorous and healthy, not inelegant, but it is difficult of cultivation, and is seldom seen thriving under artificial treatment.

COMMON PRICKLY SHIELD FERN.

This is *Polystichum aculeatum*, the *Polypodium aculeatum* and *Aspidium aculeatum* of the older writers.

FIG. 7.

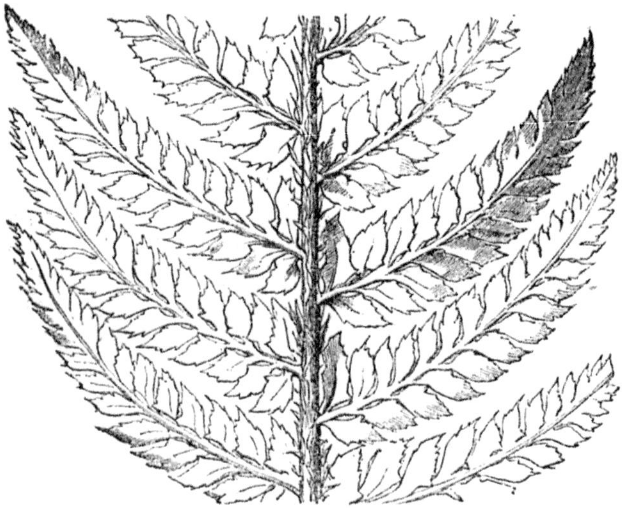

Polystichum aculeatum.

The plant is almost evergreen in a sheltered situation, and is one of those species which are well suited by boldness of character for the decoration of rocky scenery. It is a stout plant, having the fronds a couple of feet or more long, and springing from a stout tufted stem or crown, whence they grow up in a circle, about the month of April, and take a somewhat erect position. They are lance-shaped in form—

in the most perfect state of the species broadly lanceolate, but in the variety *lobatum* very narrowly lanceolate. The texture is harsh and rigid, the upper surface dark green and shining, and the short stipes densely enveloped in rust-coloured membranous pointed scales. The fronds are bipinnate, with alternate pinnæ, these pinnæ being again more or less perfectly divided into a series of pinnules, which are either decurrent, that is, insensibly merging in the substance of the rachis which supports them, or else, are tapered to a wedge-shaped base and attached to the rachis by the point of the wedge. The general form of these pinnules is somewhat elongately crescent-shaped, the upper base being extended into a small auricle, or enlarged lobe, and the lower base as it were cut away; while the apex is tapered off to an acute point, and the margin is serrated with spiny teeth. The veins are alternately branched, and do not join together or anastomose, but extend free to the margin; and the fructification, which is generally abundant, and often crowded, is ranged in a line on each side the midrib of the pinnules, and also on the larger pinnules on each side the midvein of the basal lobes or auricles. The indusium is circular, and attached by a little depression or stalk in its centre.

The variety *lobatum*, considered a distinct species by some, differs chiefly in the narrow outline of the frond, and in the pinnules being much more decidedly decurrent—that is, run together at the base; every possible variation in the consolidation of the pinnules is to be met with, between the typical bipinnate form of *Polystichum aculeatum*, and a simply pinnate form of the species, which, from its resemblance to *P. Lonchitis* has been called *lonchitidoides*. This latter form, owing its origin to the peculiar circumstances of growth only, cannot be considered as a permanent variety, but the intermediate state—that is, the plant called *lobatum*—which is the most common of these aberrant forms, is at least sufficiently different to be noticed as a variety.

This common and free-growing Fern is found in hedge-banks, and similar situations; and being abundant, easily cultivated, nearly evergreen, and withal possessing considerable elegance of growth, has much to recommend its admission to a prominent position in the Fern garden.

SOFT PRICKLY SHIELD FERN. [*Plate* IV.]

The Soft Prickly Shield Fern, sometimes called the Angular-lobed Shield Fern, is, in scientific language, called *Polystichum angulare*. It has many synonymous names, among which *Polypodium angulare*, *Aspidium angulare*, and *Polystichum setiferum*, are more commonly in use than the rest. It is a strong-growing, tufted stemmed species, sometimes forming large masses. The fronds are lanceolate, from two to four or five feet high, persistent through ordinary winters and in sheltered situations retaining their verdure unimpaired until the new fronds are produced. It is one of the most graceful of all the native species. The stipes, which varies from a third to a fourth of the length of the entire frond, is very shaggy, with reddish chaffy scales, which scales, though of smaller size, are continued throughout the upper parts of the frond. The fronds are bipinnate, with numerous tapering, distinct pinnæ, having their pinnules flat, somewhat crescent-shaped, from the prominent auricle at the anterior base, often bluntish at the apex, but sometimes acute, always with spinulose marginal serratures, and sometimes, in a few of the lower pinnules, with deep lobes, so that the pinnules become pinnatifid. The pinnules are tapered to a broad-angled base, the lines of which usually exceed a right-angle, and they are attached to the rachis of the pinna by a short, distinct, slender stalk, which does not form a line with either margin. The pinnules have branched free veins; and the sori are generally ranged in a row on either side the midrib, and are covered by a peltate scale or indusium.

The highly developed form of this species alluded to, as having its basal pinnules deeply lobed, is the variety *subtripinnatum*. It does not differ in any other particular, but, being rather more lax than the other forms, is one of the most elegant of them all. It is not uncommon. The variety *decompositum* is still more divided in the same way.

The variety *proliferum* is another very elegant and highly developed form. This has the pinnules narrowed and attenuated, more or less lobed; and the stipes bears above ground little buds, which become young plants. It has been found in Devonshire.

The variety *imbricatum* is a very remarkable form, differing from the type, in the very narrow linear-lanceolate outline of the frond, as well as in having the pinnules, which are roundish-oblong, so closely placed that they overlap each other. It also bears young plants on the stipes above the surface of the soil. It was found in Somersetshire.

The variety *alatum* is another exceedingly curious form. In this the fronds are small; and the pinnules are connected by a very obvious leafy expansion which margins the rachis, orming along the side of the latter what is technically called a wing. This is also a Somersetshire variety.

The variety *cristatum* is one of much beauty. The point of the frond, and the points of all the pinnæ, are expanded into tassel-like tufts, as occurs in the tasselled or crested varieties of the Male Fern and Lady Fern.

There are many other variations. Sometimes the pinnules are acute, sometimes blunt and rounded. In other forms the pinnules are deeply serrated, and in some they are very conspicuously spinulose. In some very elegant forms the pinnæ and the pinnules are exceedingly irregular in size and form. The variations are, however, so numerous, that we can only refer for an account of them to our *Handbook*, previously mentioned.

This normal form of this species is not an uncommon Fern; it grows in hedge-banks and in lowland woods, preferring, like most of the larger Ferns, the presence of plenty of free (not stagnant) water.

As a cultivated plant, either for pots or rockwork, the Soft Prickly Shield Fern is a most desirable plant; and acquiring, as it does, considerable size, it may be made to produce some striking effects in ornamental scenery.

CHAPTER VIII.

THE BUCKLER FERNS.

We have already mentioned that the old genus *Aspidium*, to which the English name Shield Fern was applied, has been in modern times broken up into two groups. One of these, the genus *Polystichum*, to which the name Shield Fern is retained, formed the subject of our last chapter. The other group is called *Lastrea*, and we distinguish them by the popular name, Buckler Fern.

The Buckler Ferns are known from the Shield Ferns by having the indusium, or seed-cover, round in outline with a notch at the hinder part, thus becoming kidney-shaped, and by having these covers attached to the frond by the notched part. This group includes some of the largest and most common of our native species. They are nearly all of them remarkable for their elegance, and several of them retain their fronds through the winter in sheltered situations; but with one exception, they are not strictly evergreen, and in exposed situations become bare during winter.

The estimated number of British *Lastreas* varies according to the value put upon certain differences in the plants, by different authors; we admit seven to the rank of species.

The name has been applied in honour of M. Delastre, a zealous French botanist and microscopical observer.

MARSH BUCKLER FERN.

This Fern is the *Lastrea Thelypteris* of botanists; and has, among other synonyms, those of *Aspidium Thelypteris*, *Polypodium Thelypteris*, *Polystichum Thelypteris*, *Thelypteris palustris*, and *Hemestheum Thelypteris*.

FIG. 8.

Lastrea Thelypteris.

This plant is called the Marsh Fern from its growing in marshes and boggy situations. It has a slender, extensively creeping stem, which is usually smooth and of a dark colour, producing matted fibrous roots. The annual fronds produced about May and perishing in the autumn, usually grow about a foot high, the fertile ones taller: but sometimes, when the plants are vigorous, they reach the height of two or three feet. They are of a delicate pale green, lanceolate, pinnate, the pinnæ mostly opposite, and pinnatifidly divided into numerous entire rounded lobes. The lobes in the fertile fronds appear narrower and more pointed than those of the barren, on account of their margin being rolled in. The venation of the lobes of this Fern consists of a distinct, somewhat wavy midvein, from which alternate usually forked venules branch out, and both branches bear a sorus half way between the margin and the midvein. The sori often become confluent, and are partially concealed by the bent-back margin. The indusium, or cover of the spore cases, is, in this species, small, thin, and shapeless, and is soon thrown off and lost.

The Marsh Buckler Fern has a wide geographical range, and in England and Wales occurs in numerous localities; in Scotland and Ireland it is rather uncommon.

It is not a very attractive species for cultivation, but grows freely enough if plentifully supplied with moisture, and allowed room to spread.

FIG. 9.

Lastrea montana.

MOUNTAIN BUCKLER FERN.

This Fern is named *Lastrea montana*, or perhaps more commonly *L. Oreopteris*; it has also the synonymous names *Aspidium Oreopteris*, *Polypodium Oreopteris*, *Polypodium montanum*, *Polystichum montanum*, *Lastrea montana*, *Hemestheum montanum*, and *Phegopteris Oreopteris*.

This is a very elegant species, growing shuttlecock fashion around the central crown which terminates the stem, to the height of from two to three feet; and it is, moreover, so fragrant, when drawn through the hand, as to be recognised from its kindred by this circumstance alone. The fragrance is due to the presence of numerous minute glandular bodies on the lower surface, which being bruised when the plant is handled, give out strongly a peculiar balsamic fragrance by no means disagreeable, accompanied by the peculiar starchy odour which many Ferns possess. The fronds are annual, springing up about May, and enduring through the summer. These are erect, lance-shaped in their outline, pinnately divided the stipes is unusually short, the leafy part being continued nearly down to the ground, and the lower pinnæ are so short that the frond tapers downwards as much or perhaps more than it does towards the point. The pinnæ generally stand opposite, and are narrow, tapering, and pinnatifidly divided, bearing their fructification almost close to the margins of the segments, and generally very abundantly. Each segment or lobe has a distinct and slightly sinuous midvein, which is alternately branched, the branches simple or divided, and bearing the spore-cases in clusters near their extremity.

This plant is found most luxuriant in woods, but occurs profusely on mountainous heaths. It may be considered common in England, Wales, and Scotland, in the latter country often very profuse on the mountain-sides; but in Ireland is much more rare.

As a garden plant, the Mountain Buckler Fern is effective for shady rockwork, and when established, grows freely if kept sufficiently moist.

MALE FERN, OR COMMON BUCKLER FERN. [*Plate* III.]

This is the *Lastrea Filix-mas.* Other names are *Aspidium Filix-mas, Polypodium Filix-mas, Polystichum Filix-mas, Drycopteris Filix-mas,* and *Lophodium Filix-mas.*

The Male Fern is so called from its robust appearance, in contrast with the more delicate, though similar, Lady Fern, or *Filix-fœmina.* It is one of the species which grow up annually, the fronds being destroyed by the frosts of winter, unless the situation be very sheltered, when the old fronds often remain green until the young ones are produced in spring. It is a robust-growing plant, producing its fronds in a tuft around a central crown, and when vigorous and perfectly developed is a very striking object, though its ornamental qualities are often unheeded, on account of its commonness. The stipes is densely scaly. The fronds average about a couple of feet in height, and are of a broad lance-shaped figure. In division they are what is called bipinnate, though less decidedly so than occurs in some other species; for here those pinnules only which are nearest to the main rachis are separate from each other. The pinnæ are narrow and tapering, with a few of the lowest pinnules distinct, the rest united at the base; these pinnules are of an obtusely oblong form, and serrated on the margin. The fructification of this plant is generally very copious, and is usually confined to the lower half of the pinnules, where it is crowded.

This is one of the best species to study with the view of understanding the fructification of Ferns; for here the indusium, a very important organ, is remarkably evident in fronds which have about reached their full development. In that state the indusium is as yet closed over the cluster of spore-cases, and will be seen to consist of a lead-coloured, tumid, kidney-shaped, conspicuous scale, which, at the proper time, becomes elevated on one side, to allow the dispersion of the spores. This may readily be noticed by watching the progress of the fronds after they have reached the stage just adverted to; or if they are gathered in that state for preservation in the herbarium, they are almost certain to burst, more or less, in the process of drying, before they yield up their vitality. These covers are at first little white scales.

The veins of this species are also readily seen, and each pinnule will be found to have a flexuous midvein, with alternate venules, which are simple or forked, or sometimes three-branched in different parts of the pinnule, the three branched ones, if present, occurring at the base, and the unbranched ones at the apex. The sori are borne on the branch towards the apex of the pinnule, and form a line of dots at a little distance on each side of the midvein.

The variety *incisa* of our *Handbook of British Ferns*, named *Lastrea erosa*, *L. Filix-mas erosa*, and *L. affinis* by others, is a magnificent Fern, much larger than the commoner form of the plant, growing four or five feet in height, and having the same general features as those already described, but larger in every part; the pinnules more elongated and tapering towards the point, and more deeply cut along the margin; the branches of the venules more numerous; and the sori produced over a larger proportion of the surface of the pinnule, usually reaching almost to its apex.

The variety *paleacea*, by some called *Borreri*, is remarkable for the abundant and usually golden-coloured scales which clothe its stipes and rachis; and for its blunt pinnules, and its inflexed indusium.

The variety *pumila* has the pinnules changed into small rounded lobes, and the fructification reduced to a single row of spore-cases on each side the rib of the pinna.

The variety *cristata* is a very curious and very handsome plant; it has the points of the frond and of the pinnæ dilated into a fringe or tassel, which is a very curious transformation, and is quite constant. There are several other varieties known.

The Male Fern is found abundantly all over the country in shady situations; the larger varieties are met with here and there in similar places; the other varieties are rare.

This is one of the most easy of Ferns to cultivate, and is very suitable for cool, shady rockwork, or for shady walks in woody scenery.

RIGID BUCKLER FERN.

This Fern is called by botanists *Lastrea rigida*. It has also been named, at various times, *Polypodium rigidum*, *Aspidium rigidum*, *Polystichum rigidum*, and *Lophodium rigidum*.

This very elegant Fern is of moderate size, growing upright or spreading, and from one to two feet in height. The fronds issue from the crown of a comparatively thick stem, and are annual in their duration, greeting the approach of summer with the fresh green of youth, and shrinking dead and shrivelled from the icy touch of winter. The fronds are narrowly triangular, and bipinnate, with narrow tapering pinnæ and oblong blunt pinnules, which are cut into broad rounded segments, again notched into a varying number of pointed but not spinulose teeth. The stipes is densely scaly. The veining is very similar to that of the large variety of *Filix-mas;* the pinnules having a flexuous midvein, with alternate venules again pinnately branched. The clusters of spore-cases are borne on the lowest anterior branch of each venule, that is, on the lowest veinlet on the side towards the apex of the pinnule, and they are covered by a kidney-shaped indusium, which does not fall away. Over the fronds are scattered numerous small sessile glands, which, when slightly bruised, give out a faint and not unpleasant odour.

This Fern seems confined to the limestone districts of the north of England, growing at considerable elevations. It was first found at Ingleborough, in Yorkshire, and has been since met with on the limestone ranges of Westmoreland and Lancashire.

In cultivation, this is usually a free-growing plant, more lax than in the wild state, and among the more elegant of the larger kinds.

CRESTED BUCKLER FERN.

This is the *Lastrea cristata* of botanists; and has besides received the following among other names, *Polypodium cristatum*, *Aspidium cristatum*, *Polystichum cristatum*, *Dryopteris cristata*, and *Lophodium Callipteris*.

This Fern is not a very elegant plant, but is of considerable interest on account of its rarity. It forms a thick creeping stem or root-stock, from which a limited number of narrow, very upright fronds arise early in May, and attain the average height of a couple of feet. The fronds are destroyed in autumn by the frosts. Their outline is linear-oblong; that is, from a narrow width at the base the margins run nearly parallel almost to the apex, where they narrow to the point; they are supported by a longish stipes, which is proportionally stout, and maintains this proportion upwards through the leafy portion of the frond; on the lower part it has a few scales, which are blunt ovate, membranous, and of a uniform light brown colour. The pinnæ are elongate-triangular in their outline, the broadest occurring at the base of the frond, the upper ones becoming gradually narrower, but all of the same general form,

FIG. 10.

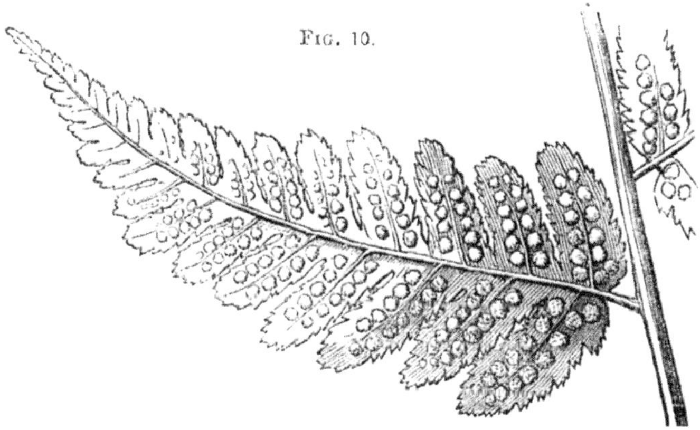

Lastrea cristata.

namely, widest at the base, gradually tapering to the apex. They are not divided quite down to their midrib, so as to become, in technical terms, pinnate, but each segment is attached by the entire width of its base, and connected by a narrow extension of its base with the segment next behind it; all the segments having their apices inclined rather towards the apex of the pinnæ. The lobes of the pinnæ are themselves oblong, with a rounded apex, and a crenately toothed margin. The midvein of the lobes takes a tortuous course, and gives off lateral branches which divide into several secondary branches, one only of which, that nearest the apex of the lobe, bears a sorus. The fructification is confined to the upper portion of the frond, and often remarkably so; less frequently it extends downwards to the pair of pinnæ next above the basal ones. The spots of spore-cases are covered by a kidney-shaped scale or indusium, having an entire margin, and become mature in August and September.

The variety *uliginosa* is exactly intermediate in its general appearance and characters between the normal form of the species and the variety *spinulosa*. It forms a stout creeping crown or root-stock, having a tendency to multiply by lateral offshoots. The stipes has ovate pallid scales. The fronds grow nearly erect to the height of from two to three feet; these erect fronds bear the fructification. Other fronds, however, are produced, which are barren, and these do not grow so erect, nor put on the same form as the fertile ones; but closely resemble those of *cristata* itself, the fertile ones having much the appearance of those of the var. *spinulosa*, only that they are narrower, and have narrow pinnæ. The outline of these latter is narrow lance-shaped, the pinnæ having a narrow tapering form, and the pinnules being oblong-pointed, with rather deep, serrated, marginal notches, the serratures terminating in a fine point. The sori are covered by even-margined, kidney-shaped scales or indusia. The barren fronds are broader, usually shorter, less erect, and their pinnules are of a broader, blunter form, and more closely placed, than those which are fertile. Sometimes after the growth of the first set of fertile fronds others spring up which are also fertile, but have the appearance described above as peculiar to the barren ones.

The variety *spinulosa* is also an erect-growing kind, with a stout creeping stem or root-stock, which becomes branched, so that several crowns are generally found forming one mass. The stipes is sparingly furnished with semi-transparent scales of a bluntly ovate form. The fronds grow from one to three feet high, and are bipinnate, the pinnæ having an obliquely tapering form from the inferior pinnules being larger than the superior ones; this is most obvious at the base of the fronds, where the pinnæ are broader than they are towards the apex. The lower pinnules on the basal pinnæ are of an oblong form, somewhat narrowing upwards, the margins deeply incised, the lobes being serrated, and the teeth somewhat spinulose.

This species occurs only on boggy heaths, and that in but few places in Britain, confined, we believe, to the counties of Nottinghamshire, Cheshire, Norfolk, and Suffolk. The var. *uliginosa* is usually found in company with it; while *spinulosa* is more abundant, but always found in marshy places and damp woods.

FIG. 11.

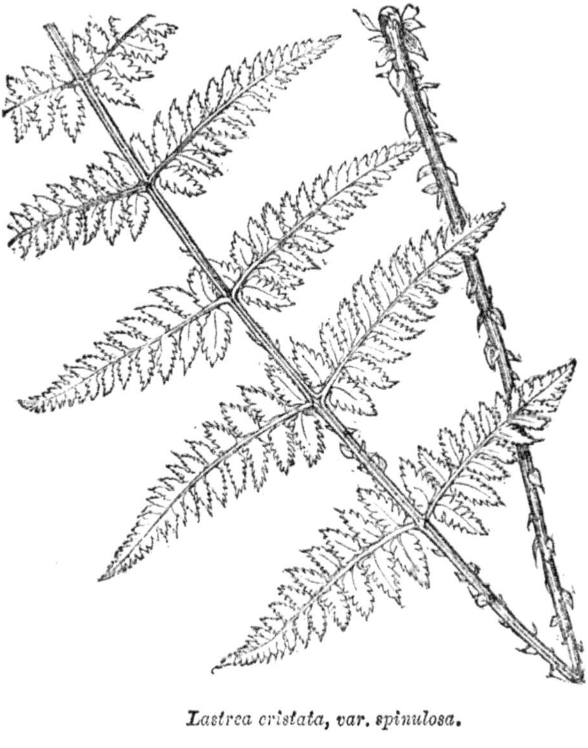

Lastrea cristata, var. spinulosa.

This Fern and its varieties are very easily cultivated on damp shady banks or rockwork, and they prefer peaty soil. When grown in pots, they require to be plentifully supplied with water.

BROAD BUCKLER FERN.

This species is named *Lastrea dilatata.* It was called *Aspidium cristatum* by some of the older botanists, and has since received numerous names, amongst which occur *Lastrea multiflora, Polystichum multiflorum, Lophodium multiflorum, Aspidium dilatatum, Polypodium dilatatum,* and *Dryopteris dilatata.*

This is one of the most compound of our native Ferns. It forms a large tufted stock or stem, and has broad arched fronds, which average about a couple of feet in height, though it is sometimes met with smaller, and often, when luxuriant, reaches a height of five feet. They are almost always more or less drooping or curved, and seldom grow erect, as those of *cristata*, *uliginosa*, and *spinulosa* do. The general outline is ovate-lanceolate, though in this, one of the most variable of Ferns, the form varies considerably, becoming sometimes narrow elongate lanceolate on the one hand, and short broad almost triangular on the other. The more usual form has the fronds ovate, lance-shaped in outline, on a stipes of moderate length, much thickened at the base, and densely clothed with entire, lance-shaped, pointed scales, which are dark brown in the centre, but nearly transparent at the margins. They are bipinnate, with elongate-triangular or tapering pinnæ, placed nearly opposite, and having more or less of obliquity from the larger development of the lower side. The pinnæ are pinnate, and the pinnules near their base often so deeply divided as to be again almost pinnate; the rest are pinnatifid, or in the upper parts merely deeply toothed; but the margins, whether deeply or shallowly lobed, are set with teeth which end in short spinous points. The veining is very similar to the more compound parts of the allied species; and the fructification is produced in great abundance. The sori are ranged in two lines crosswise the pinnæ on the larger lobes, or lengthwise on the less divided parts, and are covered by kidney-shaped scales or indusia, which are fringed around the margin with projecting glandular bodies.

One of the varieties of this Fern has the fronds shorter, almost triangular in outline, and often remarkably convex; it has, moreover, usually a dark green colour, often with a brownish tinge. It is found in more exposed places than the normal form, and is not uncommon. Another, sometimes called *nana*, seems chiefly remarkable for its small size, seldom exceeding six or eight inches in height, which peculiarity it is said to maintain under cultivation. It is rather rare, or at least local in its occurrence.

The variety *collina*, which has been called *Lastrea collina*, has the fronds ovate, drawn out to a long narrow point, or narrow oblong-lanceolate, and the pinnules, which are obtusely ovate and have a broad attachment at the base, have the serratures on their margin broader and less spinulose than in the common form. It has narrow scales with a darker centre. It was first noticed on the hills of Westmoreland.

The variety *glandulosa* is of larger growth, its surface covered with glands, and the scales of its stipes broader and paler, so that it approaches the *spinulosa* form of *L. cristata.* This has been called *Lastrea glandulosa*; and was originally found in the Forest of Dean.

The variety *dumetorum* is comparatively small, with oblong-ovate or ovate-triangular fronds, covered with glands, the stipes clothed with narrow, pointed, pale-coloured scales. It occurs on the hills of Westmoreland and Wales, and what seems the same plant, from the Isle of Arran, has been called *L. maculata.*

A more detailed account of these, and several other variations of this species, will be found in our *Handbook of British Ferns*, and in *The Octavo Nature-printed British Ferns.*

The common forms of this species, though found in drier places than *spinulosa*, are nevertheless partial to moisture, being found in damp, shady hedgebanks and woodlands.

This is a hardy Fern, and easily cultivated in almost any soil.

HAY-SCENTED, OR TRIANGULAR BUCKLER FERN.

The scientific name of this Fern is *Lastrea æmula.* The plant has, however, had many others, of which the best known are *Lastrea fœnisecii*, *Lastrea recurva*, *Lophodium recurvum*, *Nephrodium fœnisecii*, and *Aspidium recurvum.*

FIG. 12.

Lastrea æmula.

The plant is of moderate size and very elegant, drooping in habit, and possessing a crisped appearance from the recurving of the margins of all the segments of the fronds. It grows from one to two feet high, a spreading circle of triangular fronds being produced from its tufted stem. The stipes is thickly clothed with small, narrow, jagged, pale-coloured scales. The fronds are bipinnate, the lowest pair of pinnæ always longer and larger than the rest, and the pinnules on the inferior side of the pinnæ larger than those on the superior side. The pinnules are of an oblong-ovate figure, and the lowest of them often divided again into a series of oblong lobes, for the most part decurrent, but sometimes slightly stalked; the margin is cut into short spinous-pointed teeth. The veins of the pinnules are alternately branched from a sinuous midvein, and these veins give off two or three alternate venules, the lowest anterior one bearing the sorus. The exact ramification of the veins depends upon the degree in which the pinnules or lobes are divided. The fructification is distributed over the whole under surface, the sori being pretty evenly distributed in two lines along each pinnule or lobe; they are covered by small reniform indusia, which have their margin uneven, and fringed with small, round, stalkless glands. The whole frond is covered with similar glandular bodies.

This Fern, which is most abundant in Ireland and the western parts of England, occurs in damp sheltered woods, and on shady banks and rocks.

It is of an elegant drooping aspect, and is cultivated without difficulty. It is the more valuable as a pot plant, from its moderate size and its evergreen character.

CHAPTER IX.

THE LADY FERN.

THE genus *Athyrium* is that to which the Lady Fern is referred. It is one of the most variable among our native Ferns, all the various forms being plants with delicate and beautiful fronds of annual duration. They vary in size from tufts of a few inches high, to plumy masses of the height of three or four feet. The texture is thin, and almost transparent, on which account the nature of the venation and of the connexion of the parts of fructification may be here very well seen and studied. These plants serve to connect the *Aspidium*-like and the *Asplenium*-like groups of Ferns, from the former of which they differ in having the sori elongate instead of round. The sori, which form short lines, are sometimes curved at the end, or even horse-shoe-shaped; and being in age short and often dilated, approaching the rounded form, the Lady Fern has, by many writers of discrimination, been placed in the old genus *Aspidium;* but if examined while young, immediately before or after the indusium has burst, its true character will readily be seen. We here have an illustration of the inconvenience which arises from the preservation as herbarium specimens, only of such as have the fructification quite mature.

The affinity of the Lady Fern is properly with the *Aspleniums.* The mark by which the *Aspleniums* and their allies are known, in addition to the elongated form of the sorus, is its position on the *side,* not the back, of the veins; the receptable being said to be lateral. The present group is distinguished by having its indusium fringed on the free margin with capillary segments, and by the frequently horse-shoe-shaped sori, while in *Asplenium,* the margin of the indusium is without the fringe, and the sori are not turned back. There is only one indigenous species of Lady Fern.

The name is derived from a Greek word signifying *opened,* the allusion being to the turned-back position into which the indusium is forced by the swelling spore-cases, bursting out as it were like an opened door.

COMMON LADY FERN. [*Plate* V.]

The Lady Fern is named *Athyrium Filix-fœmina* by botanists; other synonymous names being *Polypodium Filix-fœmina, Aspidium Filix-fœmina,* and *Asplenium Filix-fœmina.*

The Lady Fern, on account of the exquisite grace of its habit of growth, the elegance of its form, and the delicacy of its hue, claims precedence over every other British species. The plant is tufted, the caudex of the larger varieties often with age acquiring some height, and elevating the circlet of fronds on a rude pedestal a few inches in length. The fronds are developed from the summit of this stem about May or June, a score or upwards being often produced by strong old plants; they reach maturity early in the summer, during which time a few additional fronds are generally developed from the centre; and the whole of them are, under ordinary circumstances, destroyed by the autumn frosts. The fronds are lanceolate, more or less broad, bipinnate; the pinnæ lanceolate, more or less drawn out at the point, and again pinnate, though sometimes with the bases of the pinnules connected by a narrow leafy wing, but not so much so as to render them merely pinnatifid. The pinnules, however, are more or less lobed or pinnatifid, the lobes being sharply toothed in a varying manner.

From the delicate herbaceous texture of the fronds the venation is very distinct; and is seen to consist, in each pinnule, of a wavy midvein, from which proceed alternate veins, which again produce alternate venules, and on the anterior side of this series of veins, at some distance from the margin, is borne an oblong sorus. In the larger and more divided pinnules the veining is more compound. The sori are themselves oblong, a little curved, the basal ones usually hippocrepiform or horse-shoe shaped; they are covered by indusia of the same form as the sorus, and in the case of the curved, or horse-shoe shaped sori, the indusium becomes apparently almost circular with a lateral notch, in which state it resembles a *Lastrea.* On one side the indusium is fixed longitudinally to the side of the vein which forms the receptacle; its other margin, which is fringed, or split into a number of hair-like segments, becomes free. This description applies to the commoner forms of the Lady Fern, which, however, are very variable in size, according to the situation and circumstances which influence their development, sometimes scarcely exceeding a foot in height, and at other times reaching the height of four or five feet, the latter being the result of growth in a damp, shady situation, the former the consequence of a more exposed and drier locality.

The variety *rhæticum,* sometimes called *convexum,* differs in its fronds, its pinnæ, and its pinnules being narrower or narrower-looking than in the common forms. The fronds, which seldom exceed two feet in height, are erect, and their form is narrow-lanceolate; the pinnæ are taper-pointed; the pinnules set quite clear of each other, very narrow, that is, linear, with sharp points, the margins bluntly toothed, but rolled under so that very little of the toothing is seen; the sori are very often confluent. It occurs in boggy places.

The variety *latifolium,* also a very distinct and strong-growing form, differs from the common sort, in the elongate or oblong-lanceolate outline of its fronds, and in the broad, leafy, crowded development of its ovate irregularly lobed pinnules, which are deeply toothed at the margin, with the curved sori lying near the sinuses of the lobes. It was found in Westmoreland.

The variety *marinum* has rather small fronds, usually about a foot, or a foot and a half long, lanceolate, and remarkable for the manner in which they taper from their broad centre, equally towards the base and apex. These fronds have a spreading or horizontal mode of growth; their pinnules are oblong and bluntly toothed, the teeth being almost always quite simple, not two or three-notched as is usual in the other forms; they are attached closely together, at right angles with the continuously winged rachis of the pinnæ. The sori are very short, often curved in a horse-shoe form, and crowded. It was found near Aberdeen.

There are, besides, several curious monstrous varieties of considerable horticultural interest. One called *multifidum*, of which several variations have now been discovered, has the tips of all the pinnæ, as well as of the frond itself, multifid or tasselled, which gives it a very elegant appearance. Another, called *depauperatum* or *ramosum*, is smaller, with the pinnæ reduced and irregularly tasselled, and the apex of the frond more deeply split into ragged-looking tasselled lobes. Another, called *crispum*, is a dwarf tufted plant, no larger than a bunch of curled parsley, which it much resembles, its fronds being curiously branched, crisped, and tasselled. These, which are, strictly speaking, monstrosities, have retained their characteristics for many years in cultivation, and are very elegant. See *Handbook of British Ferns*.

The common Lady Fern is abundant in warm moist woods and hedgerows throughout Great Britain, and especially so in Ireland; it also occurs throughout Europe, and in Asia, Africa, and North America.

None of our native Ferns are more easily cultivated than this. A rather boggy soil suits it best, and it loves shade and moisture; indeed, these latter conditions being fulfilled, soil becomes a secondary consideration. The moisture, however, though abundant, should not be stagnant. The Lady Fern is occasionally seen planted in the mouth of a cave or recess, by water, among shady rockwork; nothing is so lovely as a finely-grown plant of it so situated. As a pot plant it requires plenty of room, both for its roots and fronds, and must be liberally watered.

CHAPTER X.

THE SPLEENWORTS.

THE Spleenworts are called *Asplenium* by botanists. The British *Aspleniums* are small evergreen Ferns, with long narrow single sori lying in the direction of the veins which traverse the fronds; and by these marks they may be known from all other indigenous Ferns, excepting the *Ceterach*, which latter is readily distinguished from them by having the back of its fronds coated with brown scales, among which the sori are hidden. They are the types of the tribe *Asplenieæ*, which consists of Ferns having the elongate masses of fructification attached along the side of the veins, and covered by an indusium of the same elongated form as the sori themselves. There are nine species of *Asplenium* indigenous to Britain, all of them small plants, interesting to the cultivators of Ferns.

The word *Asplenium* comes from the Greek *asplenon*; a name applied by old authors to some kind of Fern possessed of supposed virtues in curing diseases of the spleen.

FIG. 13.

Asplenium septentrionale.

FORKED SPLEENWORT.

This Fern is named *Asplenium septentrionale.* It has also borne the several names of *Acrostichum septentrionale*, *Scolopendrium septentrionale*, and *Amesium septentrionale.*

A rare and diminutive Fern. The habit is tufted, large masses being sometimes formed; the fronds themselves are very small, from two to four or six inches long, seldom longer, slender, dull green, with a longish stipes, which is dark purple at the base. The leafy part—if, indeed, it can here be called leafy—is of a narrow elongate lance-shaped form, split near the end into two or sometimes three alternate divisions, or in the smaller fronds into the same number of teeth; each of the divisions of the frond has its margin cut into two or more sharp-pointed teeth, the points of the larger teeth being very frequently bifid. The veins are reduced to a minimum; one vein enters each lobe, or if the frond is not lobed the stipes is continued upwards in the form of a vein; this becomes forked so as to send up one vein to each of the teeth into which the part is divided; and three or four long linear sori are produced in a very crowded manner within this small space, so that when from age the sori burst open the indusium, the spore-cases form a confluent mass over the whole under-surface.

The confluent mass of spore-cases arising from the crowded position of the sori, has led some authors to consider this plant an *Acrostichum*, the mark of which is to have the whole under-surface thus covered. Some of the sori being face to face, growing as they do from the inward side of each vein, and almost in juxtaposition, has again led other botanists to think it a *Scolopendrium*, the mark of which is to have the sori confluent in pairs face to face. If, however, the plant is examined while young, it will be seen that these resemblances are unreal, and that it is truly an *Asplenium*.

This rare Fern is found in the West of England, but more abundant in the North and in Ireland. It is found on rocks and walls.

In cultivation it requires sandy peat-soil mixed with rubbly porous matter; and in uncongenial situations the shelter of a close-frame, or bell-glass.

FIG. 14.

Asplenium germanicum.

ALTERNATE SPLEENWORT.

This is *Asplenium germanicum*, its synonyms being *Asplenium alternifolium*, *Asplenium Breynii*, *Scolopendrium alternifolium*, and *Amesium germanicum*.

It is one of the rarest of our native Ferns, and perfectly distinct from *A. Ruta-muraria*, of which some botanists have thought it to be a variety. It grows in little tufts, the fronds being from three to six inches high, sub-evergreen, narrow-linear in form, pinnate, divided into distant, alternate, wedge-shaped pinnæ, one or two of the lowest having generally a pair of very deeply divided lobes, the upper ones more and more slightly lobed, all having their upper ends toothed or notched. The fronds are quite small, and the parts narrow, which, added to their opacity, renders the venation indistinct; there is no midvein, but each pinna or lobe has a vein entering from the base, which becomes two or three times branched as it reaches the broader parts upwards, six or eight veins generally lying close together, in a narrow fan-shaped manner, in each of the larger pinnæ, the smaller ones having a proportionately less number. Two or three linear sori are produced on a pinna, and these are covered by membranous indusia, the free margin of which is entire, or slightly sinuous, but not jagged; the sori at length become confluent.

It grows, but very rarely, in Scotland, and in the Lake district; and is found, but very sparingly, in other parts of Europe.

This kind is not only rare, but one of those which does not fully yield to artificial culture. It grows tolerably freely if potted in sandy peat-soil well-drained by an admixture of rubbly matter, and kept under a bell-glass in a shaded frame, or greenhouse; but the plants are very liable to die in winter. The safeguard is, not to allow any water to lodge about their crowns, nor to keep the bell-glass too closely or too constantly over them, especially in winter.

RUE-LEAVED SPLEENWORT, OR WALL RUE. [*Plate* VI. Fig. 2.]

This is *Asplenium Ruta-muraria*, with the following among other synonyms: *Amesium Ruta-muraria*, and *Scolopendrium Ruta-muraria*.

A very diminutive plant and not very attractive, occurring abundantly on old walls, often in such situations little more than an inch high. It grows in tufts, insinuating its wiry roots, as is the case with all the mural species, into the crevices and joints of the masonry, and is not easily removed from such places in a condition suitable for planting. The fronds are numerous, of a glaucous-green, varying between one and six inches long, with a stipes about half the entire length, the leafy part usually triangular in outline, and bipinnate. The pinnæ are alternate, with rhomboidal, or roundish ovate, or obovate pinnules, sometimes wedge-shaped with the apex abruptly cut off. The more luxuriant fronds are once more divided, so as to become almost tripinnate, the pinnules being deeply pinnatifid, and the lobes formed like the ordinary pinnules. When the plants are quite young, the fronds are simple and roundish kidney-shaped. At a later stage of development, they are occasionally only once pinnate, with pinnatifid pinnæ. The upper margins of the pinnules are irregularly toothed. The veins are rather indisitnct, and there is no midvein, but a series of veins arise from the base, becoming branched in their progress towards the apex, the number of ultimate branches usually corresponding with that of the marginal teeth. Several sori are produced near the centre of the pinna, covered by indusia, which open inwardly with a jagged or irregularly sinuated margin.

A common species, confined to rocks and walls, occurring throughout Europe and in many parts of North America.

It is not difficult to cultivate in pots or on well-drained rockwork.

GREEN SPLEENWORT.

This elegant little Fern is the *Asplenium viride* of botanists.

FIG. 15.

Asplenium viride.

The plant has such a general resemblance to *A. Trichomanes* as to have been mistaken for it by casual observers. It is, however, quite distinct, and is most readily known from *A. Trichomanes* by the colour of its rachis, which is green in the upper part, while in the latter it is black throughout. It is an evergreen tufted species, producing narrow, linear, simply pinnate bright pale green fronds, ranging from two to eight or ten inches in length, supported by a short stipes, which is dark-coloured at the very base, but otherwise green, the rachis being entirely green. The pinnæ are small, generally roundish-ovate, rather tapered towards the base, and attached to the rachis by the narrowed stalk-like part, the margin being deeply crenated.

The venation is distinct: the midvein sends off alternately a series of venules, which are either simple or forked, bearing the sori on their anterior side. The sori are oblong, covered at first by membranous indusia, which are soon pushed aside; the free margin is jagged or crenate.

A native of moist, rocky, mountainous districts in England, Scotland, and Wales; occurring, also, though less frequently, in Ireland, and throughout Europe.

This Fern is not difficult to cultivate in pots in a close, damp, cold frame; or on moist, shady rockwork, if covered over by a bell-glass. If exposed, it is apt to suffer from occasional excessive wet, which often does not properly drain away; and also from the dry hot air of our summers. The object of covering it with a glass is to avoid both these casualties, and provided it is not kept too close it will then thrive well. The proper bell-glasses for these half-hardy Ferns are those with a small opening in the crown, which may be closed or not at pleasure, but, in general, is best left open. In pots it should have a gritty, porous soil.

COMMON MAIDENHAIR SPLEENWORT.

This Fern is called *Asplenium Trichomanes.* It has also had the names of *Asplenium melanocaulon,* and *Asplenium saxatile.*

FIG. 16.

Asplenium Trichomanes.

It is rather a diminutive plant, but has a very interesting appearance, from the black stipes and rachis, and the regularity with which the bright green pinnæ are disposed. It grows in tufts, naturally introducing itself into the joints of old masonry and among the crevices of rocks, and producing numerous small slender fronds, of a linear form, in its most vigorous state nearly a foot long, but generally from three to six inches. They are evergreen, simply pinnate, on a rather short stipes, which is of a purplish-black, the rachis also being of the same dark colour. The pinnæ are deep green, small and numerous, equal-sized, of a roundish-oblong figure, attached to the rachis by a stalk-like projection of their posterior base; the margin is rather entire or crenated. The pinnæ are jointed to the rachis, and when old are readily displaced, so that eventually the black rachis is left denuded among the tuft of fronds. A distinct midvein passes through each pinna, giving off on each side a series of venules bearing veinlets, the anterior of these producing the linear sorus just within the margin of the pinnæ. The sori, which in the young state are covered by thin indusia having a somewhat crenulated free margin, very frequently in a later stage become confluent, and cover the whole of the under surface.

A very rare and very curious variety of this species, named

incisum, has the pinnæ deeply pinnatifid, with linear notched segments. Another, equally rare and still more beautiful, has the ends of the fronds tasselled; this is called *cristatum.*

The species occurs rather plentifully, growing on rocks, old walls, and ruins, and less frequently on hedgerow banks. It is pretty generally distributed throughout the United Kingdom and Ireland; and also occurs throughout Europe, and in each of the other divisions of the globe.

When once established, this plant grows readily either in pots or on rockwork; but its roots being wiry, and generally inserted into the crevices of the walls or rocks on which it grows, it is sometimes found to be difficult to transplant. In general the smaller and younger plants may be removed with greater success than the larger and older ones. The newly transplanted roots should be kept rather close, if possible, for a short time; but after they are established, shade is not so essential to this species as to most other Ferns, although it grows most vigorously under the influence of shade and shelter. In a Wardian case, for which its size is suitable, it should have the upper and drier parts of the rockwork.

SEA SPLEENWORT. [*Plate* VI. Fig. I.]

The proper or scientific name of the species is *Asplenium marinum.*

It is a very handsome evergreen maritime Fern of tufted habit, with linear or linear-lanceolate fronds, usually six or eight inches long, of the deepest glossy green, with a smooth, rather short, dark brown stipes. The fronds are simply pinnate, with stalked pinnæ, connected at their base by a narrow wing which extends along the rachis; their form is either obtusely ovate or oblong, unequal at the base, the anterior base being much developed, while the posterior is, as it were, cut away, the margin being either serrated or crenated. They are of leathery texture, but the veins are nevertheless tolerably evident, each pinna having a midvein, from which venules are given off alternately on either side, these again producing a series of veinlets. The sori are produced on the anterior side of each venule, lying obliquely, and forming two rows on each side the centre; they are oblong or linear, covered by a persistent indusium, which opens along the anterior margin as the spore-cases grow towards maturity.

The chief variation to which this Fern appears subject is that of the elongation of its parts. Sometimes the pinnæ are much elongated, tapering to a narrow point; sometimes, besides being narrowed, they are auricled at the base, and deeply lobed.

This is a maritime species, occurring profusely on our south-western rocky coasts and in the Channel Isles, and extending to France and Spain, to Madeira and the Canaries.

Fig. 17.

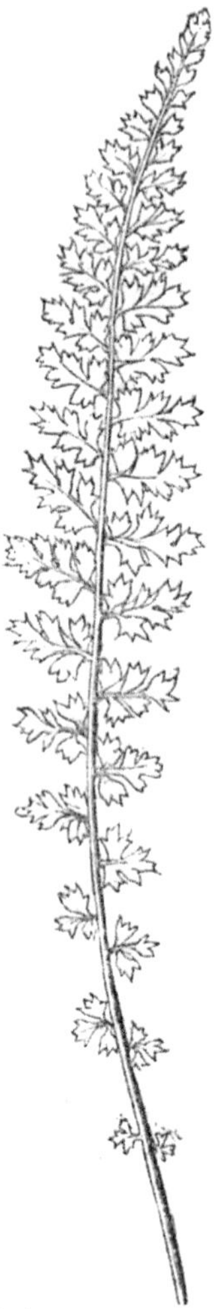

Asplenium fontanum.

In cultivation this Fern thrives most luxuriantly in the atmosphere of a damp hothouse, where it forms, in a comparatively short time, a dense mass of the deepest green, and often reaching a foot and a half in length. In a cold frame, if kept closed, well-established plants will continue in health, progressing slowly, and never acquiring half the size of those grown in heat. In the climate of London it does not prosper, nor, as far as we know, survive, if planted on exposed rockwork.

This species, with the Lanceolate Spleenwort and the Maidenhair, are exceedingly well adapted for Wardian cases in warm sitting-rooms. All of them enjoy the warmth; and being all evergreens of moderate size, and very elegant in structure, they supply just what is wanted in such situations. They should be planted on elevated rockwork, in sandy peat-soil lying in the interstices between the fragments of stone; and when once established will grow freely, provided they are not much exposed to the sun, which they do not like.

SMOOTH ROCK SPLEENWORT.

This is the *Asplenium fontanum* of botanists. Among its other names occur *Asplenium Halleri*, *Polypodium fontanum*, *Aspidium fontanum*, and *Athyrium fontanum*.

It is a small tufted-growing species, seldom seen more than three or four inches high. The small fronds are evergreen, and mostly grow nearly upright; they are of a narrow, lanceolate form, rather rigid in texture, of a deep green above, paler beneath, and supported on a very short stipes, which has a few narrow, pointed scales at the base. They are bipinnate, the pinnæ oblong-ovate, and the pinnules obovate, tapering to the base, the superior basal pinnule of each pinna having the margin divided by four or five deep, sharp teeth, the rest of the pinnules and lobes having from one to three similar teeth. The main rachis of the frond, as well as the partial rachis of each pinna, have a narrow leafy expansion along their sides, throughout their length; and this is perhaps the most obvious technical point, except size, by which to distinguish the present plant from *A. lanceolatum.* In structural details they very much resemble each other.

The fronds being rigid and opaque, the venation is often less evident than is usual in Ferns. It consists, in each pinnule, of a central or principal vein, which throws off a venule towards each lobe or serrature. On two or more of these veins a sorus is produced, which in form is short compared with those produced by most of the genus; the form being oblong, rather flat on the side by which they are attached, and covered by an indusium of similar form, which is waved and indented on the free margin. Sometimes the sori keep quite distinct, but it is not uncommon for them to become confluent so as to cover nearly the under-surface of the whole of the little pinnules.

There are some who doubt this species being really a native of Britain, on the ground that it is not now to be found in the places where it is said to have been originally met with. Considering the exact record of its discovery, and considering, moreover, that it is a very small plant, and that the places where it would be most likely to occur are generally the most inaccessible, and, therefore, the least likely to be searched; considering, further, the many probable localities which exist, and have not been carefully explored by any keen botanical eye, we are not justified in rejecting the statements which the older botanists have left us, although it has not recently been found in wild localities.

This species grows freely planted in a well-drained pot, and kept in a close cold frame; in a damp hothouse it succeeds well, becoming much more vigorous under the influence of heat.

LANCEOLATE SPLEENWORT.

This is the *Asplenium lanceolatum.* It has recently been called *Tarachia lanceolata.* An evergreen Fern of variable size. Its fronds are from four or six inches to a foot in length, bipinnate, lanceolate in form, supported on a brownish-coloured stipes on which as well as on the rachis are scattered small bristle-like scales. The more vigorous plants are nearly erect, though sometimes somewhat spreading in growth. The pinnæ spread at nearly right angles with the rachis, often opposite, and have an ovate-lanceolate form. The pinnules are of irregular form, often obovate, or nearly so, sometimes unequally quadrate, but always indented on the margin with deep, sharp teeth, the larger pinnules being first lobed, and the lobes toothed, the smaller ones simply toothed. The venation is tolerably distinct; the pinnules each having a tortuous midvein, which produces forked venules, and these produce veinlets, one of which extends towards each serrature. The sori are at first oblong, and covered by an indusium of the same form, having a lacerated free margin; but as they become old the sides become bulged out so as to give them a roundish form, and the indusium becomes obliterated.

FIG. 18.

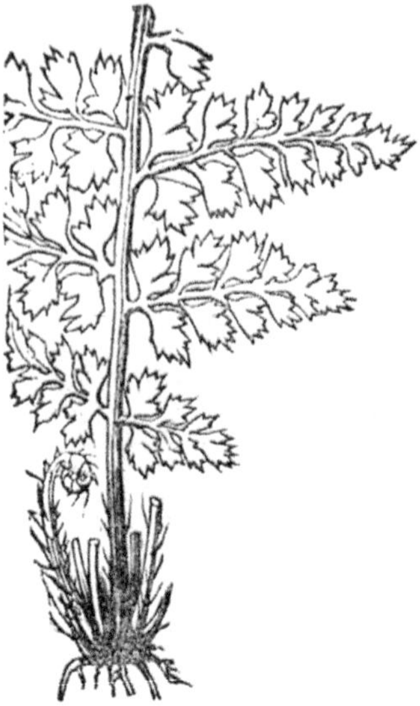

Asplenium lanceolatum.

The variety *microdon*, a very rare plant, has pinnate fronds, the pinnæ being only undulated and lobed, not again pinnate.

This is rather a local species, being found only in the southern and western parts of England, and in Wales, almost always near the coast. It is found very luxuriant in the Channel Islands.

As might be expected, it evidently requires a mild and sheltered climate, so that in a hothouse, where the temperature is not kept too high, or in a green-house, it grows freely; this cannot always be said of plants kept in a cold frame, and never of plants fully exposed, unless the locality is very favourable.

BLACK MAIDENHAIR SPLEENWORT.

The Black Maidenhair Spleenwort is the *Asplenium Adiantum-nigrum* of botanists, and has, moreover, been called *Tarachia Adiantum-nigrum*, and *Asplenium lucidum*.

FIG. 19.

Asplenium Adiantum-nigrum.

It is a rather common evergreen Fern, and a very conspicuous ornament of the situations where it occurs in a vigorous state. The fronds grow in tufts, and vary much in size, from a height of three or four inches when it occurs on walls, to a foot and a half and even two feet including the stipes, when it occurs on shady hedge-banks in congenial soil. They are triangular, more or less elongated at the point, the shining dark purple stipes being often as long as, or longer than, the leafy portion; but in stunted plants growing in sterile situations very much shorter. They grow erect or drooping, according to the situations in which they occur. They are bipinnate, or sometimes tripinnate; the pinnæ pinnate, triangular-ovate drawn out at the point, the lower pair always longer than the next above them. The pinnules, especially those on the larger pinnæ, are again pinnate; the alternate pinnules being deeply lobed, and the margins sharply serrate. Each pinnule has a distinct midvein or principal vein, bearing simple or branched venules, on which the sori are produced. At first the sori are distinct, and have the elongate narrow form common to this genus, but as they become older they often spread and become confluent, so that almost the entire under-surface of the frond is covered with the spore-cases. The indusium is narrow, with its free margin entire; this soon becomes pushed away by the growing sori, and is lost.

FIG. 20.

Asplenium Adiantum-nigrum, var. acutum.

The variety *acutum* differs, principally, in the more decidedly three-cornered fronds, which, in consequence of their shortness and breadth, and the high development of their basal pinnules, form a nearly equilateral triangle; in the very much attenuated apices of the fronds and their pinnæ, which are, in fact, what is called caudate; and in the extreme narrowness of the ultimate segments into which the very much divided frond is cut, these segments being narrow, linear, and acute. The fronds grow a foot or upwards in length, including a long brown stipes. In large specimens the leafy portion is about six inches long, and as much across the base, triangular, tripinnate. The lower pinnæ are considerably larger than the next pair, and elongately triangular. The primary pinnules are ovate-acuminate; the secondary pinnules lozenge-shaped, these latter being cut down almost to the centre into linear sharply two- to five-toothed segments. The venation consists of a vein, which enters each lobe of the pinnule, and branches alternately into as many nearly parallel venules as there are marginal teeth, one venule being directed into each tooth. The narrow linear elongate sori are borne, rather close together, on these venules. It is a very rare plant, having been found in a few Irish counties, and in Jersey. It is met with in the North of Europe, and more plentifully in the Canaries, the Azores, and Madeira.

The ordinary forms of the plant are very commonly met with growing on rocks or old walls, and on hedge-banks in a sandy soil. The latter situations, where they grow most vigorously, are often beautifully adorned by their drooping tufts. The extreme forms are more rare.

This is one of the more useful evergreen Ferns for shady rockwork, as it will grow with freedom if planted in sandy soil, which is just kept moistened either by natural or artificial means. As a pot plant it is easily manageable. The variety is rare, and has hitherto been treated as a frame or greenhouse plant.

CHAPTER XI.

THE HART'S-TONGUE FERN.

THE botanical name of the Hart's-Tongue Fern is *Scolopendrium.* The genus is botanically very distinct from all our other native Ferns; and from other points of view is exceedingly interesting. There is only one British species, but of this there are numerous varieties, which have a perfectly distinct aspect, owing to peculiarities in their development. They are all evergreen, and on this account, as well as by reason of their hardiness and bold striking appearance, they are among the most ornamental of all Ferns for out-door rock-work. The genus is known by the peculiarities of its sori, which, though forming parallel oblique lines at intervals on each side the midvein, and having the appearance of being single if seen when mature, are in reality composed of two sori, set face to face, and so close together as to become confluent along their whole length. The fructification, technically speaking, consists of sori confluent in pairs, placed face to face.

Scolopendrium is merely an alteration of *Scolopendra,* the scientific name of the centipede; and the name is applied from a fancied resemblance (in the position we suppose) between the feet of a centipede and the lines of its fructification.

COMMON HART'S-TONGUE. [*Plate* VII. hinder fig.]

The proper name of this plant is *Scolopendrium vulgare,* but many others have been given it, as *Scolopendrium officinarum, Scolopendrium Phyllitis, Asplenium Scolopendrium,* &c.

The Hart's-Tongue Fern is a common plant; nevertheless, in consequence of its shining bright green, though simple fronds, contrasting so beautifully with the feathery aspect much more common among the Ferns, it does not want for admirers whether seen in a wild or cultivated state. It grows in tufts. The fronds, which are evergreen, vary in length from six inches to a foot and a half, and even more, and are either stiff and erectish when growing under circumstances which render them dwarf, or more or less spreading and drooping when in situations which are favourable to enlarged development: in the former case the fronds are thicker and more leathery in texture; in the latter, thinner and less rigid, from being produced in very damp shady situations. The usual form of the fronds is what is called strap-shaped, that is, narrow oblong-lanceolate, much elongated; they taper towards, and are acute at, the apex, narrowing a little downwards, and becoming cordate at the base; the margin is entire, or very slightly wavy, and they are supported on shaggy stipes averaging about a third of their entire length. The fronds have a strong midrib or costa, extending throughout their whole length, from which are produced forked veins, the branches of which (venules) lie parallel, and proceed direct towards the margin, terminating just within the edge in a club-shaped apex. The sori, which are oblong patches of unequal length, lying in the direction of the veins at short intervals along the upper two-thirds of the length of the frond, are each composed of two proximate lines of fructification laterally united; each line, however, consisting of a complete sorus, so that the two united are properly called a twin sorus. The indusia which cover these, have their attachment on the upper and lower sides of their respective venules, the other edges overlapping one the other.

This is the ordinary form of *Scolopendrium;* but there are a great number of very curious and some very distinct varieties, differing only, however, in the form of the fronds, and not in the fructification, where it is present. These varieties, which are noticed at length in our *Handbook of British Ferns*, are for the most part perfectly constant under cultivation, although they have, no doubt, originated in aberrations—that is to say, accidental variations, from the original species, which have been perpetuated naturally or by art. It is moreover a curious fact, that most of them are reproduced from spores.

The variety *crispum* is one of the most beautiful of them; in this, the same outline of frond prevailing, the leafy portion is so much more developed than the midrib, that the margin becomes excessively undulated, giving the fronds a very elegant curled or crisped appearance. This sort is barren.

The variety *polyschides* is very curious and distinct. The fronds of this are linear, and blunt at the apex, much narrower than in the common sort, and the margin is deeply and irregularly lobed and crenated. This sort is fertile, and its sori are short, forming two irregular lines on each side the midrib.

The variety *marginatum* is another curious and very beautiful form, lobed in the same manner as *polyschides*, but having the fronds broader; it is remarkable also in having, behind, a longitudinal excurrent membrane on each side between the midrib and margin, on which membrane as well as exterior to it, the short interrupted sori are produced.

Another striking variety is *multifidum*. This has the fronds forked either near the apex or sometimes near the base; each branch is again more or less repeatedly forked, and the apices of all the forks are developed into irregular fan-shaped leafy expansions, to which the term multifid is applied. Sometimes the fronds are merely forked once or twice, without being multifid; in other cases the stipes itself becomes forked, bearing multifid branches, and this has been called *ramosum*. This multifid sort is fertile; and occurs in many variations.

The variety *laceratum* is a dwarf and highly ornamental form; in this the fronds are often nearly as broad as long, with the margin deeply gashed into irregular lobes, the lobes being numerous, crowded, and much undulated, sometimes tapering, sometimes more or less dilated at the apex, the basal pair often considerably enlarged, and so much developed as to produce an approach to the palmate form.

The common Hart's-tongue is an inhabitant of hedge-banks, of old walls, and sometimes of the interior of wells, in which latter situation it acquires great luxuriance. It is one of the more commonly distributed species in England and in Ireland, less abundant in Scotland; and also found all over Europe. The varieties are rare in a wild state, and are better known as cultivated plants.

An evergreen, and a plant of free growth, the Hart's-tongue is one of the most desirable hardy Ferns we possess for open rockwork. Its simple fronds contrast well with the more compound forms; and its varieties all have a different aspect, combined with the same good qualities of hardiness and endurance. Shady and rather humid places are those in which this plant most delights, although, as is evident from its sometimes growing on walls, it will live in more exposed and arid situations. The plants, however, never acquire much vigour under such circumstances, and have mostly a starved and stunted aspect. They are not particular as to soil, sandy loam containing fibrous or half-decayed vegetable matter, is much preferable to soil which is much spent and comminuted, as indeed is the case with respect to all Ferns.

CHAPTER XII.

THE SCALE FERN.

The adopted botanical name of the Scale Fern is *Ceterach.* Of this genus there is only one British species, and this plant is so different from all others as to be distinctly recognised at a glance. The mark by which it is known is this:—the back of every frond is covered by densely-packed, brown, pointed, chaffy scales. Among these scales, and concealed by them, lie the elongate sori, which are anomalous, in regard to their relationship, in having no indusium. The affinity of *Ceterach* is without doubt with the *Asplenium*-like Ferns, this being the case they ought to have an indusium. No indusium, however, exists here, unless it be represented by a kind of membranous ridge, which is to be found on the receptacles just behind the sori, and is the part which has sometimes been called an indusium. The probability is, that it does represent that organ, which is not largely developed in consequence of the presence of so dense a covering of scales, these not only serving the purpose of a cover to the sori, but perhaps, from their crowded position, preventing the proper formation of the usual form of cover.

The name *Ceterach* is an alteration of the word *Chetherak,* which was applied to this plant by Persian and Arabian medical writers.

COMMON SCALE-FERN, OR SCALY SPLEENWORT. [*Plate* VI. fig. 3.]

This species has many names. That most to be preferred is *Ceterach officinarum,* but it has also been called *Asplenium Ceterach, Scolopendrium Ceterach, Grammitis Ceterach, Notolepeum Ceterach,* and *Gymnogramma Ceterach.*

It is a dwarf, evergreen, distinct-looking and very pretty Fern, growing in tufts. The fronds when fresh are thick and fleshy, and from this cause they are perfectly opaque when dry. Their size varies, according to the circumstances of their growth, from two to six inches in length, rarely exceeding the latter. They grow on a short scaly stipes, and are either pinnatifid, as is commonly the case, or more rarely pinnate, the difference being, that in the latter the fronds are divided rather more deeply than in the former. The upper surface is a deep opaque green; and the under surface is densely covered with rust-coloured brown closely-packed overlapping scales, which, being just seen projecting from the margin, and still more fully in the exposed under surface of the young partially-developed fronds, prettily contrast with the deep green of the upper surface. The pinnæ or lobes are of an ovate form, and either entire or lobed on the margin. The opacity of the fronds renders the venation indistinct, and indeed it is only to be made out by examining young fronds, removing the covering of scales, and the outer skin of the frond itself. It is then seen, that from the lower corner the principal vein enters, taking a sinuous course towards the upper side of the apex; it branches alternately, the venules being again branched, and the veinlets becoming joined more or less near the margin. The sori are borne along the sides of the venules in a very irregular manner, the majority of them being directed towards the apex of the pinna. At first, the sori are quite concealed by the scales, but the spore-cases ultimately protrude between them.

The *Ceterach* is a mural species, occurring on the walls of old buildings and ruins, and in rocky places. It is pretty generally distributed in the United Kingdom, but is considered somewhat rare in Scotland. It occurs also throughout central and southern Europe, and in the north of Africa.

Like other wall Ferns, this is often difficult to establish in cultivation when first transplanted; but when once this is overcome its cultivation is not difficult. It is best grown in a cold frame, potted rather high, among loam mixed with a large proportion of brick-rubbish, and not over-watered.

CHAPTER XIII.

THE HARD FERN.

This plant is referred by some authors to *Blechnum*, and by others to *Lomaria*. We think it most nearly related to the former, although in the contraction of its fertile fronds it undoubtedly resembles the latter. Among the British Ferns the only species of this genus is known by having its fructification extended longitudinally on the pinnæ, so as to form a linear or continuous sorus on each side the midvein, and about midway between it and the margin. No other British

Fern has its fructification in extended lines lying parallel with the midrib, except the *Pteris*, or Bracken, in which however the sorus is on the margin, and not within the margin and near the midvein, as in *Blechnum*. The *Blechnum* may, however, be at once known from the *Pteris*, by the division of its fronds, which are merely pinnate, while those of *Pteris* are decompound.

The name *Blechnum* is an adaptation of the Greek *blechnon*, which signifies a Fern.

COMMON HARD FERN. [*Plate* VIII. right-hand fig.]

This plant is the *Blechnum Spicant*. It has also the following names: *Blechnum boreale, Lomaria Spicant, Asplenium Spicant, Onoclea Spicant, Acrostichum Spicant, Struthiopteris Spicant, Osmunda Spicant*, and *Osmunda borealis*.

The common name of this species is very appropriate, from the rigid harshness of its texture. It is one of the few native kinds which produce two distinct-looking kinds of frond—fertile and barren. The fertile ones have their pinnæ much narrowed, or contracted, as it is called, while the fronds themselves are considerably taller than the barren ones. These fronds grow in large tufts, and being very gracefully bestowed, the plant becomes one of the most ornamental of our wild species during the summer season, when its fronds are in a fresh state. Both kinds of fronds are of a narrow lanceolate form; the barren ones being only deeply pinnatifid, while the fertile ones are pinnate; but the segments in both are long and narrow, like the teeth of a comb. The barren fronds, which are from one-half to two-thirds the height of the fertile ones, assume a spreading or horizontal position, and are attached to the caudex by a very short scaly stipes. The fertile ones, which are situated in the centre of the tufts, are erect, from one to two feet high, the stipes, which is sparingly furnished with long pointed scales, being nearly half the length, and of a dark brown colour. The veins are not very evident in the fertile fronds, on account of the contraction of the parts, but they resemble those of the barren ones, except in having a longitudinal venule on each side the midvein, forming the receptacle to which the spore-cases are attached. The midvein is prominent, and produces a series of venules on each side, these becoming forked, and extending almost to the margin, terminating in a club-shaped head. In the fertile fronds the veinlets are necessarily shorter, and connected, as already mentioned, by the longitudinal venules which bear the fructification. The spore-cases are thus arranged in two linear sori, one on each side the midvein; these are distinct while young, but soon become confluent, covering the whole under-surface of the pinnæ. The indusia, by which they are at first covered, when mature, burst along that side towards the midrib, and eventually become split across here and there, at points opposite some of the venules.

The hard Fern is a rather common plant, occurring in heathy and stony places, and preferring localities which are rather damp than otherwise. It is found in various parts of Europe.

In cultivation, this is a very suitable plant for damp shady rockwork, and in such situations, planted in peaty soil, it grows freely, and without requiring any special attention.

CHAPTER XIV.

THE BRAKES, OR BRACKEN.

The *Pteris* or Bracken is the most common of all our Ferns. It is that which occurs almost everywhere in woods and in sandy wastes, often appropriating to itself the whole surface of the ground. It is variable in appearance, owing to differences in its size and development dependent on the circumstances in which it grows. Its more usual size is from three to four feet in height. Sometimes in dry, very sandy soil, the plant becomes a pigmy, not reaching a foot in height, and being merely bipinnate. The opposite extreme occurs when the plant is growing on damp hedge-banks in warm, shady lanes, where it attains eight or ten feet in height, and is proportionately compound in its development. Under circumstances which favour the most luxuriant development, this common and usually vulgar-looking plant combines the most noble and graceful aspect, perhaps, which is borne by any of our indigenous species, its fronds scrambling up among the bushes which sustain them at the base, while their graceful feathery-looking tops form overhead a living arch of the tenderest green.

The *Pteris* is known among our native Ferns by having the edges of all the little divisions of its fronds furnished with a line of spore-cases. No other of our native species has the fructification arranged in continuous lines except *Pteris* and *Blechnum;* and the *Pteris* may be readily known from that by the lines being in it confined to the margin, leaving the centre unoccupied, while in *Blechnum* the extreme margin is unoccupied by the sori.

Pteris is a Greek name for a Fern, and is derived from another Greek word, which signifies *feather;* and, of course, is applied in reference to the graceful feather-like aspect which the fronds of Ferns generally possess. When the plant is very luxuriant this name is quite as applicable to the Bracken as to any other known Fern. This consideration is perhaps enough to justify the application to this species, by the older writers, of the name of Female Fern, which scarcely seems appropriate to the commoner uncouth-looking form which the plant more usually bears.

COMMON BRAKES, OR BRACKEN. [*Plate* IX.]

The botanical name of the bracken is *Pteris aquilina;* that of *Eupteris aquilina* has also been proposed.

This Fern has a caudex that creeps very extensively beneath the surface of the soil. This caudex is thickish, black-looking, and succulent, containing a good deal of starch. From it are produced, at intervals, the annual fronds, which generally make their appearance about the latter end of May. The fronds themselves have been variously described, and often erroneously, for they are not unfrequently said to be three-branched; but except when very much starved and stunted, do not approach that form very nearly. They are, in reality, bipinnate, or when very luxuriant tripinnate, the pinnæ standing opposite in pairs, each pair in succession becoming fully developed, while the main rachis is extending upwards, and the next pair is beginning to unfold. The mature fronds are thus twice or thrice pinnate, with the pairs of pinnæ standing opposite. The stipes is downy while young, and the part under ground is black, like the creeping stem itself, and spindle-shaped just at the base. Average specimens of the fronds are tripinnate, that is, they produce a certain number of pairs of branch-like pinnæ, which branches are bipinnate. We must confine our further description to one of these branches, selected from the lower part of the frond. The general form is ovate, a little elongated; that of its pinnæ (the secondary pinnæ) narrow lanceolate. These latter are placed rather closely together, and are again divided into a series of pinnules, which are either undivided or more elongated, and deeply pinnatifid or sinuate. Each pinnule of the undivided form has a distinct midvein, producing alternate lateral venules, which become twice forked, and extend to the margin, where they meet a longitudinal marginal vein which forms the receptacle. The indusium consists of a bleached, membranous, fringed expansion of the upper skin or epidermis of the fronds, which reflexes so as to cover the spore-cases, but there is here another membrane which lies beneath the spore-cases.

This, which is the most abundant of our indigenous species, is also widely distributed in other parts of the world, and bears a variety of names, from having been supposed to be distinct by those who have met with it from such widely separated localities.

Being so common, and in an ordinary state uncouth-looking, it is not a plant for cultivation to any extent. In warm, damp, wilderness-scenery, however, where it would attain great luxuriance, and the situation is such as would enable it to develope the arching character already mentioned, it might very properly be introduced.

CHAPTER XV.

THE MAIDEN-HAIR FERN.

THE *Adiantum*, or Maiden-hair Fern, may be known by its almost fan-shaped leaflets or pinnules, which are attached by their narrow end to the little black hair-like stalks. This, however, though sufficient by which to recognise it, among the very limited number of kinds which are found in a wild state in Britain, is not its proper distinctive mark. The real characteristics lie in the veins and in the sori. The former may be readily seen by holding a pinnule between the eye and a strong light, and the latter by lifting up the little reflexed lobes which occur here and there at the margin on the under surface. The veins are dichotomously forked, that is, separating into two equal branches, beginning from the base upwards, the forking being several times repeated. The sori are produced on the reflexed (or bent under) membranous expansions of the margin of the fronds, which form the indusia, these indusia being traversed by veins which bear the sori. There is only one native species which possesses these characteristics, and this is certainly one of the most beautiful, as it is also one of the rarer of our Ferns a and being of small size and of evergreen habit, it is one of the most desirable of all for culture in a Wardian case.

The name of the genus comes from a Greek word, which signifies *dry*, or *unmoistened*, and is applicable to these plants, from their possessing in a remarkable degree the property of repelling water.

COMMON MAIDEN-HAIR FERN. [*Plate* VII. front fig.]

The common Maiden-hair Fern is the *Adiantum Capillus-Veneris* of botanists.

It is a small evergreen species, furnished with a very short creeping stem, which is clothed with small black scales, and bears delicate, graceful, somewhat drooping fronds, of six inches to a foot high. These fronds are usually of an irregularly ovate form, sometimes elongate, occasionally approaching to linear. The fronds are twice or thrice pinnate; with alternate pinnæ and pinnules. The ultimate pinnules or leaflets are very irregular in shape, but for the most part have a wedge-shaped or tapering base, and a more or less rounded and oblique apex, and they have generally some variation of a fan-shaped or rhomboidal outline. The margin is more or less deeply lobed, the apices of the lobes in the fertile pinnules being reflexed and changed into membranous indusia, whilst the lobes of the barren fronds are serrated; their texture is thin and membranaceous, their surface smooth, their colour a cheerful green. The stipes, which is about half as long as the frond, and furnished with a few small scales at the base, is black and shining, as also are the rachides, the ultimate ramifications of which are small and hair-like.

The veins throughout the pinnules are forked on a dichotomous or two-branched plan, from the base upwards. The sori are oblong, covered by indusia of the same form, each consisting of the apex of one of the lobes of the frond, changed to a membranous texture, and folded under.

The Maiden-hair is a local plant, though it has a wide geographical range. It is found here and there in the warmer parts of Great Britain and Ireland, evidently preferring cavernous and rocky situations within the influence of the sea. The same species is found in the warmer parts of Europe, in Asia, in the north of Africa, and in the Canaries and Cape de Verd Islands.

It is, moreover, a tender plant, and does not thrive under cultivation in the climate even of the south of England, unless sheltered in a frame or greenhouse, or by being covered with a glass. In a Wardian case it grows well; and attains great luxuriance in a damp hothouse. The proper soil for it is very light turfy peat, mixed with a considerable proportion of silver sand, and it is beneficial to plant it on or around a small lump of free sandstone.

CHAPTER XVI.

THE BLADDER FERNS.

THE botanical name of this group is *Cystopteris*. The species of *Cystopteris* are all small, fragile Ferns, yet, notwithstanding, they are very beautiful and very interesting. They are much more delicate and herbaceous in their texture than the majority of our native species, and hence are well adapted for the purpose of minute investigation into the nature of their venation and fructification. Their texture alone almost suffices to tell a practised eye their family position, but the tyro needs a more precise characteristic, and this is found in the structure of the scale or indusium which covers the sori. The sori of these plants are round, as in *Lastrea* and *Polystichum*, all, equally with *Cystopteris*, once included under the old family name of *Aspidium*; but here, instead of being almost flat and circular, the cover is inflated or bulged out like a hood, and is attached at the back (towards the base of the pinnule) of the sorus by its broad base, covering the spore-cases while in a young state, but becoming ultimately reflexed at the point, which is more or less jagged or fringed. Hence these plants are called Bladder Ferns. There are three native species, of one of which numerous distinct forms or varieties occur.

The technical name comes from two Greek words, which respectively mean *bladder* and *fern*; so that in this case the English appellation is a literal translation of the scientific name.

BRITTLE BLADDER FERN. [*Plate* x. fig. 2.]

This Fern, generally known among botanists as *Cystopteris fragilis*, has a host of other names. Some of these are *Cyathea fragilis*, *C. cynapifolia*, *C. anthriscifolia*, *C. dentata*; *Cystea fragilis*, *C. angustata*, *C. dentata*; *Polypodium fragile*, *P. cynapifolium*, *P. anthriscifolium*, *P. dentatum*. *P. rhæticum*; *Aspidium fragile*, *A. dentatum*, and *A. rhæticum*.

The Brittle Bladder Fern is a tufted-growing plant, spreading, if undisturbed under congenial circumstances, into large patches of numerous crowns; each of which throws up a tuft of several fronds, growing from six inches to a foot, sometimes more, in height. The stipes is very brittle and shining, with a few small scales at the base. The frond is lanceolate, bipinnate, the pinnæ lanceolate, the pinnules ovate acute, cut more or less deeply on the margin, the lobes furnished with a few pointed teeth. In some of the plants, and usually owing to their vigour, the pinnules are so very deeply cut as to become pinnatifid, almost pinnate, the lobes themselves then resembling the smaller pinnules nearer the apex of the pinnæ and frond. The venation in ordinary-sized pinnules consists of a somewhat tortuous midvein, which gives off a lateral branch or venule to each of the lobes into which the margin is cut, these venules branching again into two, three, four, or more veinlets, according to the size of the lobes, and each branch generally bearing a sorus at about midway its length. The sori are thus generally numerous, and rather irregularly disposed; and it often occurs that they are so numerous as, when fully grown, to become confluent into a mass of fructification covering the whole under surface of the frond.

FIG. 21.

Cystopteris fragilis, var. Dickieana.

The sori, which are nearly circular, are covered while young by a concave or hood-shaped indusium, which usually becomes torn or split at the point into narrow segments, and the whole soon becomes pushed back or cast off by the growing spore-cases.

The variety *dentata* is generally smaller, and almost always blunter in the form of its parts; this grows from six to eight inches high, and has ovate-lanceolate pinnæ, with ovate, obtuse, pointless pinnules, which are again divided on the margin into a series of short blunt notches or teeth; the venation is more simple, and the fructification is more marginal, than in any of the preceding forms. It is reproduced from the spores.

The variety *Dickieana* is of a more compact habit than the preceding, and grows from four to six inches in height; the outline almost ovate, terminating in a point; the pinnæ ovate-lanceolate, deflexed, overlapping each other; the pinnules decurrent, broad, obtuse, with a few shallow, marginal notches; the texture very delicate and herbaceous; and the fructification marginal. It is of a deep green. It is a constant variety under cultivation, and is reproduced by spores.

FIG. 22.

Cystopteris regia.

The usual forms of this species occur abundantly in moist mountainous districts, and also on walls, but generally in moist rocky situations throughout the United Kingdom, Ireland excepted, where it is comparatively rare. The same species is very widely dispersed in various parts of the world. The varieties are more rare. *Cystopteris fragilis* may be said to have rather a preference to limestone.

Under cultivation it is one of the most manageable of the smaller sorts, growing freely on rockwork or in pots. Its fronds are produced very early in spring, are often renewed during summer, and continue to grow up in succession until the frosts cut them off.

ALPINE BLADDER FERN.

The name of this species is *Cystopteris regia. Cystopteris alpina* is another name for this elegant plant, which has also been called *Cyathea regia* and *Cyathea incisa, Cystea regia, Polypodium regium, Polypodium alpinum, Aspidium regium,* and *Polypodium trifidum.*

This diminutive but elegant plant is quite a gem. It has

a close tufted stem, producing from its crown numerous bright green fronds, usually four to six, but sometimes as much as ten inches high. These grow up in May, and die away in autumn. Their form is lanceolate, the mode of division bipinnate, with the pinnules so deeply pinnatifid as to render them almost tripinnate. The stipes is short, smooth, and scaly at the base. The pinnæ are nearly opposite, with a winged rachis, ovate, divided into bluntly ovate pinnules, these latter being deeply cleft, almost down to their midvein, into short, blunt, linear lobes, which are either entire, or have two or three blunt teeth. The midvein of the pinnules is nearly straight, with a venule, simple or divided, branching off to each lobe, one branch extending to the point of each marginal tooth. The small roundish sori are rather numerous, but not confluent, borne near the margin, and covered by a concave membranous indusium.

This species, which may be cultivated without difficulty, in pots, under shelter, provided they are guarded against the effects of damp in winter, has been found on an old wall at Leyton, in Essex. It occurs in the alpine parts of southern Europe.

MOUNTAIN BLADDER FERN.

The mountain Bladder Fern is *Cystopteris montana.* Its synonyms are *Polypodium montanum, Aspidium montanum, Cyathea montana, Cystopteris Allioni,* and *Cystopteris myrrhidifolium.*

This is the rarest of our native Ferns. It is a small species, growing with a slender creeping scaly stem. The fronds are from four to six or eight inches high, triangular in outline, from the great development of the lowest pair of pinnæ; tripinnate in the lower part, and bipinnate upwards, the pinnæ spreading, and standing opposite in pairs, the lowest pair considerably larger than the next above, and unequally developed, the inferior side being very much larger than the superior. The lower pinnæ, on the inferior side, are first divided into ovate or lanceolate pinnules, and these are again cut into a second series of pinnules, of an ovate or oblong form, these ultimate pinnules being coarsely and irregularly notched or toothed; on the upper side, the pinnules correspond with the secondary pinnules of the lower side. The inferior pinnules of the next pair of pinnæ also correspond in size, outline, and subdivision with the secondary pinnules of the lower pinnæ; and above this the parts become gradually smaller and less divided up to the apex of the frond. The whole texture of the frond is delicate and herbaceous, as in the more common species, and hence the veins show very distinctly; in the ultimate pinnules the central vein is somewhat flexuous, and gives off alternate lateral veins, one of which is directed towards the sinus or marginal indentation between two serratures. The sori

FIG. 23.

Cystopteris montana.

have the roundish form common in this genus, and, being often numerous, they then become very conspicuous when full-grown; but though crowded they do not appear often to become confluent. These sori are covered, in the young state, by a blunt, concave, jagged-edged indusium.

This rare species occurs only among the Breadalbane mountains of Scotland, and in the Clova Mountains in Forfarshire. In the European Alps it is met with, most abundantly northwards; and it also occurs on the Rocky Mountains of the New World. It grows in very wet shady places on the ledges of rocks.

CHAPTER XVII.

THE WOODSIAS.

The *Woodsias* form a family group consisting of two diminutive kinds, which, however, possess much interest among the British species on account of their extreme rarity. These Ferns are furnished with indusia, and by the peculiar construction and position of this organ, they may readily be known. The peculiar nature of the indusia consists in their being placed not as a cover to the sori, but attached underneath them; when very young they indeed enclose them, but subsequently they split from above into narrow scale-like segments not easily distinguished, without optical assistance, from the hairs which occur along with them on the fronds. In the full-grown state, the sori are consequently seated in the centre of a spreading tuft of hair-like scales, which are formed of the lacerated margins of the indusium—the latter being attached to the frond at the point beneath the capsules. No other native Ferns possess a structure at all approaching to this.

The name *Woodsia* was given in compliment to a veteran English botanist, Joseph Woods, Esq., author of a very useful *Tourist's Flora.*

OBLONG WOODSIA. [*Plate* VIII. left-hand fig.]

This is the *Woodsia ilvensis.* It has been called *Acrostichum ilvense,* and *Polypodium ilvense.*

This Fern is a deciduous species, dying down to the ground annually in winter, and reviving with the returning spring. Its very short stems form tufted masses. The fronds average about four inches in height, and are less frequently found larger than smaller than this. Their form is lanceolate, more or less broad; they are pinnate, the pinnæ usually set on nearly or quite opposite in pairs, and having an obtusely oblong outline, with a deeply-lobed or pinnatifid margin. They are more or less clothed on both surfaces, but especially on the veins beneath, with minute bristle-like scales, and shining jointed hairs, among which the sori are almost concealed. The stipes is also scaly, and, as occurs in a whole group of these *Woodsias*, has a joint or articulation at a short distance from its base, at which point separation takes place if the fronds are left on to attain a good old age, the lower part remaining attached to the caudex, while the upper part falls away. The veining of the segments of the pinnæ consists of a rather indistinct midvein, from which the venules, either simple or branched, proceed towards the margin, near to which the sori are produced.

ALPINE WOODSIA.*

The name of this species is *Woodsia alpina*. It has also been called *W. hyperborea*. Their rarity rather than their beauty invests these plants with interest for the cultivator. They require to be kept in a cold shady frame, to be potted in porous soil amongst lumps of stone, to be carefully guarded against drought or stagnant moisture, and to be rarely disturbed at the root.

FIG. 24.

Woodsia alpina.

This is a diminutive species, never exceeding a few inches in stature, and renewing its fronds annually in the spring, the older ones being destroyed by the frosts and cold of winter; the fronds are cast off at the articulation or joint near the base of the stipes, which occurs in this family. The plant grows in a tufted manner, sending up several fronds from the crown. The fronds are longish and narrow; they are pinnately divided into several roundish triangular pinnæ, which are shallowly lobed on the margin, and are usually set on alternately along the opposite sides of the stalk or rachis; those towards the lower part are usually placed at a greater distance apart than those near the upper end. They are nearly smooth on the surface, and, in this respect, unlike those of the kindred species, which have a much more hairy appearance; small hair-like scales, in company with hairs, are however present in this species. The midvein of the pinnæ is indistinct, and throws out venules into each lobe, these venules being more or less branched according to the size of the lobes. The sori are placed near the extremity of the venules, and are often abundantly produced, so as to become crowded on the pinnæ.

The two species of *Woodsia* are found only in high mountain regions, where they grow from the crevices of the moistened rocks. They are both rare, though, from the inaccessible localities in which they only occur, they may really be more abundant than is generally supposed.

CHAPTER XVIII.

THE BRISTLE FERN.

THE *Trichomanes*, or Bristle Fern, is one of the most rare among our native Ferns; the one indigenous species being among the few which are very seldom met with, and that within a very narrow range. Unlike in texture all the other native kinds excepting the *Hymenophyllums*, being quite pellucid, and of the most delicately crisped appearance imaginable, it may be distinguished from them by this mark alone. The fructification, too, is here totally unlike that of all others, except the *Hymenophyllums*. The technical mark distinguishing *Trichomanes* and *Hymenophyllum* from the other British Ferns, is found in their spore-cases being contained within deep urn-shaped pits or recesses at the margin; the fructification therefore being at the *margin* instead of at the *back* of the fronds. *Trichomanes* is known from *Hymenophyllum* by its urns, or involucres as they are called, being entire, while those of *Hymenophyllum* are split lengthwise into two valves. In both, the spore-cases are clustered around hair-like receptacles, which are the ends of the veins of the fronds projecting into the urns. In *Trichomanes* it is usual for these receptacles to project more or less, so that the fronds become somewhat bristly when very full of fructification; and hence has arisen the common name of Bristle Fern, which is applied to the group.

The name *Trichomanes* itself has the same signification: it comes from two Greek words, meaning *hair*, and *excess*, in reference to these projecting hair-like bodies.

EUROPEAN BRISTLE FERN. [*Plate* X. fig. 1.]

This Fern, like many others, has had many names. *Trichomanes radicans* is here adopted; but more or less in use will be found *Trichomanes speciosum* and *Trichomanes brevisetum*.

and among the more ancient names *Trichomanes alatum* and *Hymenophyllum alatum* occur.

This very beautiful plant exists only in the immediate neighbourhood of waterfalls, and in situations where a constant moisture is maintained. Such conditions are, indeed, quite necessary to it, on account of its semi-membranous texture, which shrinks before an arid atmosphere; and hence it can only be successfully cultivated when kept quite close, and constantly wetted over-head. This species has a creeping, wiry, black-looking stem, clothed with scales. The fronds are three or four times pinnatifid, cut up into small linear segments, which are entire or bifid at the apex, and have a stout nerve or vein running up their centre, and rendered very conspicuous in consequence of the thin pellucid texture of the leafy expansions which surround it. Or the frond may be described as consisting of a series of three or four times branched rigid veins, margined throughout by a thin, pellucid, cellular expansion or wing, a greater or less number of the apices of the veins becoming surrounded by the cellular membrane in the form of an urn or vase, and within them bearing the fructification

The fronds are pendulous, and vary from an angular-ovate to a lanceolate form, the divisions being considerably undulated, so that they acquire a crisped appearance. The first series of lobes are usually of an ovate-lanceolate form; the next series shorter, more ovate, and the third series of divisions narrow, more or less linear. The ultimate branches of the veins which extend into the divisions of this third series, end just at or within the apex of the lobes if they are barren; but if they are fertile, they are produced beyond the margin, and surrounded at the base by the urn-shaped involucre, within which the spore-cases are placed. Sometimes the involucre is so placed as to appear immersed within the margin, but it more frequently projects. The length of the bristle-like receptacle is variable.

The variety *Andrewsii* is the lanceolate form of this plant, in which the pinnæ or first set of lobes are narrow and distinct.

The Sister Isle now claims, so far as the British Isles are concerned, sole parentage of this lovely, half-transparent species; there, amidst dripping rocks, it thrives with a degree of luxuriance which charms every one who has seen it creeping over their shelving ledges. It is said to have been formerly found in Yorkshire. The same species is widely distributed in the warmer parts of the world.

The variety and elegance of this plant make it a favourite species for cultivation. The conditions of success are, a close atmosphere, shade, moderate warmth, constant but not stagnant moisture, and a porous surface to which the roots may cling.

CHAPTER XIX.

THE FILM FERNS.

THE British *Hymenophyllums*, or Film Ferns, are small moss-like plants, with pellucid fronds, distinguished, along with *Trichomanes*, by having their fructification at the edges of the fronds; and known from that genus by having the involucres which surround the clusters of spore-cases, two-valved instead of urn-shaped or entire. They are the smallest of all our native Ferns, and, being somewhat rare, or at least local in their distribution, they have always been regarded with much interest. Two native species are recognised, much like each other in general aspect, and distinguished by one or two rather minute technicalities, which, however, are sufficiently obvious to those who have learned how to look for them.

The name *Hymenophyllum* is compounded from the two Greek words which mean a *membrane*, and a *leaf*; and is applied to those plants with much propriety, from the membranous texture of their leaves or fronds.

TUNBRIDGE FILM FERN. [*Plate* XII. fig. 2.]

The name of this species is *Hymenophyllum tunbridgense*, the *Trichomanes tunbridgense* of older writers, so named in consequence of its having been found in the neigbourhood of Tunbridge, though occurring also in many other parts of the United Kingdom.

It grows in the form of matted tufts, on the surface of damp rocks, in the sheltered humid localities which are congenial to it; the black, wire-like, creeping stems being entangled together, and interlaced with the mosses and allied plants which are often found in its company. The fronds are very short, from one to three or six inches long, membranous and semitransparent, almost erect, and of a dull brownish-green even when fresh, which gives them in some measure the appearance of being dead. These fronds are lanceolate, or somewhat ovate; they are pinnate, with the pinnæ pinnatifid or bipinnatifid, and having their branches mostly produced on the upper side, though sometimes alternately on each side the pinna. The fronds are virtually, as is the case with the *Trichomanes*, a branched series of rigid veins, winged throughout, except on the lower part of the short stipes, by a narrow, membranous, leafy margin. The clusters of spore-cases are produced around the axis of a vein, which is continued beyond the margin of the fronds, this vein or receptacle being enclosed within an urn-shaped involucre, consisting of two nearly orbicular compressed valves, which are spinosely serrate on the upper margin.

This species is widely distributed throughout the United Kingdom, and is found in many other parts of the world.

The *Hymenophyllums* require the same conditions for their successful cultivation as does the *Trichomanes*, to which genus the reader is referred.

WILSON'S FILM FERN.

The name of this species is *Hymenophyllum unilaterale*. It is perhaps more commonly known as *H. Wilsoni*.

It is a small moss-like plant, with numerous creeping filiform stems, generally growing in dense tufts, and producing a crowded mass of semi-drooping, brown-green, half-transparent fronds, averaging three or four inches in height. The fronds are of a linear-lanceolate form, and pinnate; the rachis is usually somewhat curved, and the pinnæ are convex above, all turned one way, so that the fronds become more or less unilateral; the outline of the pinnæ is wedge-shaped, cut in a digitate-pinnatifid way, the lobes being linear-obtuse with a spinulose-serrate margin. The rigid veins, branching from the principal rachis, which is very slightly winged in the upper part, become themselves branched so as to produce one venule to each segment; or, in other words, the veins are twice branched, and throughout their entire length after they leave the central rib they are furnished with a narrow

FIG. 25.

Hymenophyllum unilaterale.

membranous leafy wing or border, the rib itself being almost quite without any such border. The clusters of spore-cases are collected around the free ends of veins, which usually occupy the place of the lowest forward segment, and are included within an urceolate involucre, which is divided into two oblong convex inflected valves, which are quite entire at the flattened edges where they meet.

This kind of Film Fern is equally diffused with the allied species; indeed, it seems to be the more common of the two in some parts of Scotland, and in Ireland. It is widely distributed in other parts of the world.

CHAPTER XX.

THE ROYAL FERN.

The *Osmunda* is called the Royal Fern, and well it deserves the regal honours, for it is the most majestic of our indigenous Ferns. It is known by its large size, by having its fronds entirely leafy in the lower part, and entirely fertile at the top, the pinnæ or branches at the apex of the fronds being changed from the ordinary leafy form, into dense masses of spore-cases, arranged in the aggregate in the same way as the leafy pinnules would have been. This mode of bearing the fructification renders it so strikingly obvious at first sight, and gives the plant an aspect so entirely different from that of those in which the fructification is more or less concealed by its position on the under surface, that the *Osmunda*, though one of what are classified as flowerless plants, is often anomalously called the Flowering Fern. In truth, the contracted chocolate-coloured apex looks not unlike a dense panicle of small brown flowers crowning the tall straight stem, whose lower pinnæ have much the appearance of broad green leaves. There is but one native species.

The name of the genus has given rise to some speculation. Some derive it from the Saxon *mund*, which they say signifies strength. Others consider the word expressive of domestic peace, and derive it from the Saxon *os*, house, and *mund*, peace. Others, again, have thought it commemorative, as the following legend sets forth:—"At Loch Tyne dwelt the waterman Osmund. Fairest among maidens was the daughter of Osmund. Her light brown hair and glowing cheek told of her Saxon origin, and her light steps bounded over the green turf like a young fawn in his native glades. Often, in the stillness of a summer's even, did the mother and her fair-haired child sit beside the lake, to watch the dripping and the flashing of the father's oars, as he skimmed right merrily towards them over the deep blue waters. Sounds, as of hasty steps, were heard one day, and presently a company of fugitives told with breathless haste that the cruel Danes were making way towards the ferry. Osmund heard them with fear. Suddenly the shouts of furious men came remotely on the ear. The fugitives rushed on. Osmund stood for a moment; then snatching up his oars he rowed his trembling wife and fair child to a small island covered with the great Osmund Royal, and helping them to land, bade them to lie down beneath the tall Ferns. Scarcely had the ferryman returned to his cottage, than a company of Danes rushed in; but they hurt him not, for they knew he could do them service. During the day and night did Osmund row backwards and forwards, ferrying troops of those fierce men. When the last company was put on shore, Osmund kneeling beside the bank, returned heartfelt thanks to heaven for the preservation of his wife and child. Often in after years did Osmund speak of that day's peril; and his fair child, grown up to womanhood, called the tall Fern by her father's name."

OSMUND ROYAL, OR FLOWERING FERN. [*Plate* XI.]

The scientific name of this noble Fern is *Osmunda regalis.* This plant has a very stately aspect, growing to the average height of three or four feet, but sometimes found eight or ten feet high. The stem by degrees acquires height, so that in very old and luxuriant plants there is a trunk formed from a foot to two feet high. From the crown of the trunk, whether seated close to the ground, or elevated, grow the fronds. When young the fronds have generally a reddish stipes, and a glaucous surface. These fronds, which are annual, growing up in spring, and perishing in the autumn, are lanceolate, bipinnate; the pinnæ lanceolate or ovate-lanceolate, with pinnules of an oblong-ovate form, somewhat auricled at the base, bluntish at the apex, and saw-edged along the margin. Some fronds are entirely barren, and these differ from the fertile ones only in having the leafy pinnules continued all the way to the apex, instead of having the apex contracted, and bearing the spore-cases. The usual condition of the fructification is, that a few of the shortened pinnæ, which form the apex of the frond, are contracted and soriferous throughout.

The venation, as seen in the barren fronds, consists of a prominent midvein, bearing once or twice forked venules proceeding to the margin in direct lines. In the fertile parts of the frond, only the midrib of the pinnules is fully developed, and the spore-cases are attached to a small portion of the venules which becomes developed just to serve as a receptacle. The spore-cases are subglobose, shortly stalked, reticulated, and two-valved, opening vertically.

The Osmund Royal is a widely-distributed plant, occurring in favourable localities, that is, marshy and boggy situations, throughout the United Kingdom, and, as already mentioned, extremely abundant and luxuriant in some parts of Ireland. It is common throughout Europe.

This plant is especially suited, in cultivation, to occupy the base of rockwork abutting upon a piece of water, where its roots may be placed within the reach of the water. It should have peat earth for its roots. The best way to establish it is, to procure strong vigorous patches from localities where it abounds, and these, if removed carefully, will succeed perfectly.

CHAPTER XXI.

THE MOONWORT.

THE *Botrychium* or Moonwort is a small and very distinct plant, easily known by two circumstances,—first, it has two fronds or rather two branches of its frond, the one of which is leafy, the other seed-bearing; and secondly, the pinnæ of the leafy branch are crescent-shaped, with the outer margin jagged. There is no other native plant which has these peculiar features, and hence the Moonwort is a plant very easily recognised when it is met with. There is another peculiarity in this Fern which also serves to distinguish it, and its near ally the *Ophioglossum*, from all other native species—the venation is straight, not circinate; that is, the fronds, before they are developed, are not rolled up spirally, unrolling as they expand, but in the incipient state the parts are merely folded together by a flat surface. Only one species of *Botrychium* is indigenous.

The name of the genus is derived from a Greek word signifying *a cluster*.

COMMON MOONWORT. [*Plate* XII. fig. 3.]

The usual name of the Moonwort is *Botrychium Lunaria.* It was formerly called *Osmunda Lunaria.*

This is a very peculiar, almost stemless plant, furnished with a few coarse brittle fibres, and a bud springing from the permanent point which represents the stem. The new fronds spring up annually, and perish before winter, and in the majority of cases are not very conspicuous; they vary from three to eight or ten inches in height, the lower half consisting of a smooth, hollow stipes; above, the frond is separated into two branches, one of which is spreading, pinnate, leafy, oblong; the pinnæ crescent-shaped, or somewhat fan-shaped approaching to lunate, filled with a radiating series of forked veins. The other branch is erect, fertile, divided into branches corresponding with the pinnæ, and these into another series of branchlets, on which, distinct, but clustered, the globose stalkless spore-cases are produced. The spore-cases are two-valved, and open transversely when ripe; the valves are concave.

This species is widely distributed, but local, occurring in open heaths and pasture, where the soil is peaty or sandy, and not wet. The same plant occurs in other parts of Europe, and also in North America.

The Moonwort is not very easily cultivated. It may, however, be preserved in pots in a cold frame, if transplanted while dormant, or when just starting, into peaty or sandy loamy soil, and kept from either of the extremes of drought or saturation. The roots should not be disturbed when once established.

CHAPTER XXII.

THE ADDER'S TONGUE.

THE *Ophioglossum* or Adder's Tongue is very nearly related to the Moonwort, though at first sight having a very different aspect. The points in which it agrees are, that the parts are folded up straight in the young undeveloped state, and the fronds are two-branched, one branch being leafy, the other fertile. It differs most obviously in its parts being all simple, while those of *Botrychium* are compound. Its habit of growth is precisely the same, but the fructification is very different, consisting of a two-ranked spike of imbedded spore-cases. There are but two native species.

The name *Ophioglossum* literally means Adder's-tongue, which is the English name borne by this plant. It is derived from Greek words which mean *a serpent,* and *a tongue.*

COMMON ADDER'S TONGUE. [*Plate* XII. fig. 1.]

This is the *Ophioglossum vulgatum* of botanists.

It is a small stemless plant, producing a few coarse brittle roots from a central crown which annually produces a bud from which, about May, a new frond arises. The fronds grow from six to ten or twelve inches in height, with a smooth, round, succulent stipes below, and becoming divided, in the upper part, into two branches, the one of which is leafy, entire, ovate-obtuse, traversed by veins which form elongated meshes. The fertile branch is erect, contracted, about half its length being soriferous, forming a linear slightly tapering spike of two lines of crowded imbedded spore-cases. The spore-cases are considered as being produced on the margins of a contracted frond; when mature, the margin splits across at intervals corresponding with the centre of each spore-case, so that eventually the spike resembles a double row of gaping roundish cavities.

The Adder's-tongue is very abundant in the localities where it is found, which are damp meadows and pastures, on a loamy soil. It is generally distributed over England, but is less abundant in the other parts of the United Kingdom. The species is a common European plant.

There is no difficulty in cultivating the Adder's-tongue, whether in pots, or among an out-door collection of Ferns; the essentials are a stiff loamy soil, and the constant presence of water enough to prevent drought.

DWARF ADDER'S TONGUE.

The Dwarf Adder's Tongue is the *Ophioglossum lusitanicum* of botanists.

This species is technically distinguished by the small lanceolate fleshy barren branch of its fronds; and by its being altogether much smaller than the common species. From the crown of the rhizome the frond rises to the height of about one and a half to three inches, and is divided above into a barren leafy branch, and a spicate fertile branch. Occasionally a barren radical frond, of lanceolate form, accompanies the two-branched frond. The barren branch is spreading, lanceolate, narrowing towards, but bluntish at the apex, and tapering at the base; from three-fourths of an inch to an inch and a half long, thick when fresh, so that the slender veins are not seen; they are, however united in very much elongated meshes. The fertile branch or spike is somewhat taller than the barren branch, supported by a footstalk, which is thickened upwards; itself about half an inch long, linear, with a tapering apex, and bearing along each margin about six imbedded spore-cases, which at length burst transversely.

FIG. 26.

Ophioglossum lusitanicum.

The existence of this curious little plant in Guernsey, was first made known in 1854. One remarkable feature of the plant is the very early period of the year at which its growth is made. By the middle of January it is fully developed, and the fronds no doubt perish early in the spring. The range of this *Ophioglossum* appears to be extensive; for it is recorded to inhabit the sandy coasts, both of Europe and Africa, washed by the Mediterranean Sea; and to extend to the Canary Islands and Madeira. It is not improbable that a diligent search might be rewarded by its discovery in the western counties of England, or in Ireland. Its early development and speedy decay should, however, be borne in mind by those who may undertake the search.

CHAPTER XXIII.

THE BRITISH CLUB-MOSSES.

The Club-mosses are, as the name implies, moss-like plants mostly of creeping or prostrate habit; with slender fork-branched stems, which are throughout their whole length clothed with leaves so placed as to overlie each other like the tiling of a roof. The fructification is produced in the axils of some of these leaves, in most of the species confined to those at the apex of the branches, where it forms a cone-like head. The organs of reproduction at once distinguish the Club-mosses from all other plants.

The family group of the Club-mosses consists of two genera, or less comprehensive family groups, which are technically called *Lycopodium* and *Selaginella*. The true *Lycopodiums* are known by having kidney-shaped spore-cases, containing minute powdery or granular spores; these have been called antheridia. In the *Selaginellas*, an additional kind of spore-case is produced, which contains three or four roundish fleshy spores, many times as large as the granular spores just mentioned, and marked at the apex by three elevated ridges; these larger bodies are called oophoridia. The true explanation of these parts is a matter of doubt; all that seems certainly known being, that the larger spores or oophoridia, germinate, or at least vegetate. It has been usual to regard both sets of organs, when present, as axillary to the leaves or bracts, and so they may be considered for all practical purposes; but a different theoretical explanation has been given of them.

These plants, like the Ferns, are most abundant in hot, humid, and especially insular situations in the tropics, becoming scarcer northwards, but often even in very northerly regions covering large tracts of land. Our native species, with one exception, are found most abundantly on the high lands of the north, decreasing in quantity as they advance southwards. Many of the tropical Club-mosses and *Selaginellas* are extremely beautiful: some are of scandent habit, and many of them attain considerable size.

Though of humble growth, and altogether unattractive in appearance, the Club-mosses are not without their use. More than one species is used in dyeing operations, and several have a medicinal reputation. The powdery spores, often called pollen, produced in considerable quantities by one common species, are highly inflammable, and used in pyrotechny under the name of vegetable brimstone. Being of a drying and healing nature, this 'pollen' is also used to prevent excoriation; and in pharmacy is used sometimes for coating pills, as it is with difficulty wetted. The Common Club-moss is emetic, and the Fir Club-moss is a cathartic and a powerful irritant; the former is used in the treatment of cutaneous disorders, and is a reputed remedy for a dreadful disease called *Plica Polonica*.

The species of Club-mosses now existing have been thought to be the direct representatives of the great tree-like *Lepidodendra* of a former age met with in a fossil state, which must have rivalled our coniferous trees. The evidence in support of this view has been questioned; but there seems no good reason to doubt, at least, that there is a very close affinity between the two races; and, indeed, some of the most skilful investigators of this subject find an almost complete agreement between them.

The British species are with one exception included in the genus *Lycopodium*, the name of which comes from *lycos*, a wolf, and *podos*, a foot, and is given in allusion to the supposed resemblance of its forked fertile stems to the claw of some animal, as of the wolf. Hence one species, and that which probably suggested the name, has been called Wolf's-claw. The name *Selaginella* is a diminutive of *Selago*, the specific title of one of the common species of Club-moss.

THE FIR CLUB-MOSS.

This is the *Lycopodium Selago* of botanists. It is one of the commonest of the species, and is usually of upright growth, the others being decumbent. This upright habit, which is evidently natural to it, often, however, gives way before the force of gravity, and in such cases the lower part of the stems is found to be somewhat recumbent, while the upper parts retain an upright position. The stems vary from three or four to six or eight inches high, and are branched two or three times in a two-forked manner; they are stout, tough, rigid, nearly of equal length, producing a level-topped tuft, and thickly clothed with imbricated leaves which are arranged in eight rows. These leaves are lance-shaped, acute, shining green, leathery in texture, and smooth on the margin; in plants which have grown in exposed places they are shorter and more closely pressed to the stem; while in plants developed in more confined and humid situations they are longer, less rigid, and more spreading.

The fructification is in this species not borne in terminal spikes as in the other kinds, but is produced in the axils of the leaves at the upper part of the stems. The spore-cases are rather large, sessile, kidney-shaped, two-valved, and filled with minute pale yellow spores.

Besides the ordinary spores, the plant is furnished with other means of propagation in the shape of deciduous buds, produced for the most part in the axils of the leaves, about the apices of the branches. These buds separate spontaneously, fall to the ground, and there vegetate, first producing roots, and then elongating into a leafy stem. They are formed by an altered leaf, which, becoming somewhat swollen on the outside, protrudes from its inner margin five smaller lanceolate leaves or teeth, the whole being elevated on a short hardened footstalk. Within this is a whorl of five parts representing a gemma or bud; the three inner lobes of this series are large and prominent, and of an ovate oblong acute form; the two outer lobes are very small, scale-like, one closely appressed to the anterior, the other to the posterior surface of the bud. In the centre of the three inner lobes, in due time, appears a thickish oblong body, which is in reality the undeveloped stem, and eventually elongates, puts out small leaflets, and becomes a plant.

These buds are capable of growth either while attached to their parent stem or when detached and in contact with the soil; and they appear to be the chief means of propagation possessed by this species, for the statements which have been made respecting the germination of the spores of the Fir Club-moss are open to much doubt. Probably it was these buds which were caused to germinate.

There is no doubt this plant possesses some medicinal properties, though it is not now used in regular practice. It is powerfully irritant, and is used by country-people, in the form of an ointment, as a counter-irritant in parts near the eye, for diseases of that organ; it appears to be also sometimes employed as an emetic and cathartic, but not without danger. A decoction is, on the authority of Linnæus, used in Sweden to destroy vermin on cattle. It is also employed for dyeing, and to fix the colour of woollen cloths.

THE INTERRUPTED CLUB-MOSS.

The *Lycopodium annotinum* of botanists. A very distinct plant, easily recognised by the interrupted leafing of its stems, the leaves being at distant intervals much diminished in size and less spreading in their direction, these points indicating where the annual growths have commenced and terminated. It is known by its narrow leaves spreading out from the stem on all sides, and arranged in five indistinct rows. It is a large-growing species, often a foot high, with irregularly branched stems, which, after they have produced fruit-spikes, or have reached an equivalent age, become depressed, rooting and throwing up another series of upright branches. The annual increase of the stems is well marked by the closer-pressed and shorter leaves which occur at the upper part of each growth, and this is what gives the interrupted appearance to the stems. The leaves, which do not decay for several years, are linear-lanceolate in form, and have their margins minutely serrulate, or finely saw-toothed, and their apex drawn out and terminating in a rigid point; they are attached directly to the stems without stalks, and are arranged in an indistinctly spiral or somewhat five-ranked order. The lower leaves, that is to say, those remaining on the older portions of the stem, are more spreading than those on the younger growth, and indeed on the oldest portions often become somewhat bent back.

The spike of fructification is in this species perfectly stalkless, being seated directly on the termination of the leafy branch. It is about an inch long, oblong, consisting of closely overlapping bracts, of a roundish-ovate form, having a long narrow point and jagged membranous margins. In the axil of these bracts is produced a large reniform capsule, containing numerous minute pale yellowish spores. The bracts become reflexed when these spores have escaped from the burst capsule.

This is a rare species, confined to wild mountainous localities, occurring in the Scottish Highlands and the Northern Isles, and in Carnarvonshire and the Lake district. It is plentiful in the pine-forests of the north of Europe, and in some parts of North America.

FIG. 27.

Lycopodium clavatum.

THE COMMON CLUB-MOSS.

This is the *Lycopodium clavatum* of botanists. It is of procumbent habit, having vigorous creeping stems often many feet in length, much branched, and attached to the soil here and there by means of tough pale-coloured wiry-looking roots. The young branches, which are very thickly clothed with leaves, grow rather upwards at first, but soon all become prostrate, and cross and interlace, forming a close matted tuft, whence comes, in fact, the name it bears in Sweden—*Matte-grass*, or mat-grass. The stems are densely clothed with small, narrow lanceolate, flattish leaves, which remain fresh through the winter; they are smooth on the margin, and terminate in a long white point. The upright stalks supporting the spikes are bare of leaves, but have at intervals whorls of similar smaller bodies called bracts, closely pressed to the stalk.

The spikes of fructification are usually over an inch in length, and are supported by a stalk of twice or more their own length. They are commonly produced in pairs, though sometimes singly, and occasionally three together on the same stalk. These spikes are cylindrical, erect, consisting of crowded triangular-ovate acuminate bracts of a pale yellow colour, having membranous serrated margins. In the axils of these bracts the spore-cases are produced, and these are subreniform, two-valved, and filled with innumerable sulphur-coloured powdery spores. The bracts become reflexed after the spore-cases have shed their contents.

This is a common species, growing in moors and heathy places in mountainous and hilly tracts of country throughout England, Wales, and Scotland; and frequent, though less abundant, in Ireland.

The leafy stems of this species are used for dyeing purposes, as well as to fix colours in the stead of alum. The long slender stems, used under the name of Stag's-horn Moss, are formed into pretty ornaments for the houses of rustics, and for decorating their fire-places during summer. Linnæus relates that in Lapland the boys have their heads decorated with chaplets formed of it, which—the twin spikes projecting on all sides—have the effect of calling up the idea of groups of fauns and satyrs. Indeed, the long flexible stems are not badly adapted for various decorative purposes.

THE MARSH CLUB-MOSS.

This is the *Lycopodium inundatum* of the botanists; a diminutive and common plant, very frequent on moist heaths and commons in the southern parts of England, less common northwards, comparatively rare in Wales and Scotland, and not found in Ireland. It prefers to grow on spots from which the turf has been pared, and is of prostrate habit, with simple stems two or three inches long, growing close to the surface of the ground, to which they are firmly attached by a few short stout roots. They are thickly clothed with narrow linear-lanceolate acute-pointed entire leaves; those on the barren horizontal stems are curved upwards. The plant extends itself at the point, throughout the growing season, the other end meanwhile undergoing a process of decay, so that in winter, when the growth is arrested, the decay still going on, the living stem is much reduced, and a small portion only remains over to produce new foliage the following season.

The spike of fructification, which is produced towards autumn, is seated at the top of an erect branch, clothed throughout with leaves of the same shape as those on the horizontal stems; the branch and the spike nearly of equal thickness throughout, the spike about an inch long, the branch rather more. The spike is green, and is formed of narrow linear-lanceolate bracts, rather dilated at the base, and sometimes having one or two shallow teeth on each side. The spore-cases are in the axils of these bracts.

THE SAVIN-LEAVED CLUB-MOSS.

This, the *Lycopodium alpinum* of botanical writers, gets its trivial name from the resemblance between its branches clothed with the closely-pressed leaves, and those of the Savin, *Juniperus Sabina.* It is a pretty little evergreen plant, forming thick wide-spreading patches of round, tough, creeping, sparingly leafy stems, bearing numerous other erect stems, which are repeatedly fork-branched, growing erect, from three to six inches high. The branches are set with small smooth sessile leaves, whose form is lance-shaped, ending in a point; and on the lower ones these leaves are more closely placed, but arranged in four tolerably regular lines, so as to give a squarish form to the branches. The little tufts of branches are for the most part level-topped, those which bear spikes of fructification being however longer than the barren ones.

The fructifications consist of little spikes, terminating a portion of the branches, erect, close, cylindrical, yellowish-green, and sessile on the branches, that is, joined to the leafy portion below, without any intermediate stalk-like part. The spike consists of a number of bracts closely packed together, each having in its axil a capsule containing numerous minute pale yellowish spores. The bracts become reflexed after the spores have been dispersed. The plants are firmly fixed to the soil, by means of tough strong wiry branched roots, produced at intervals along the prostrate stems.

The head-quarters of this species is in elevated mountainous tracts. It occurs very abundantly in Scotland and Wales; in the northern isles; on the hills of the north, and extending into the south-west of England. It is less common in Ireland. It also occurs throughout the Alpine districts of Europe and Northern Asia.

The Savin-leaved Club-moss is a bitter plant, with a somewhat aromatic flavour, and possesses emetic properties; it is, however, seldom applied to any use. According to Sir W. J. Hooker, it is used in Iceland as a dye for woollen cloths, to which it gives a pale and pleasing but not brilliant yellow. The process is simply that of boiling the cloth in water, along with a quantity of the *Lycopodium*, and some leaves of the Bog Whortleberry.

THE PRICKLY MOUNTAIN MOSS.

This is the *Selaginella spinosa* of scientific botanists, though it is probably more generally known by the name of *Lycopodium selaginoides*, which it formerly bore. It has a slender, procumbent, often branched stem, the barren branches short and wavy, the fertile ones ascending or erect, and from two to three inches high. They are clothed with lance-shaped leaves, of a delicate texture, jagged along the margins with spiny teeth; those on the decumbent stems being shorter, as well as more distant and spreading, than those of the fertile branches.

The inflorescence, as in the other species, is a terminal spike of about an inch in length, consisting of lance-shaped jagged-edged bracts, larger and more closely pressed than the leaves of the stem. These bracts protect two kinds of fructification; the lower ones bear in their axils large three-celled spore-cases containing three globular oophoridia, and the upper ones bear subreniform spore-cases, containing the minute pulverulent pollen-like spores. This is the only native *Lycopod* which produces the two separate kinds of spores.

Though hardly to be considered rare, this is one of the less common species. It is found in the north of England, Wales, and Scotland, in which latter country it is pretty generally distributed. In Ireland it is rather common. The localities which it prefers are wet boggy places by the side of mountain rills.

The *Lycopodiums* are not frequently seen in cultivation, but they nevertheless, equally with the Ferns, would become a source of much interest if brought constantly under the eye in a living state; and in an equal degree the study of them in this condition the watching of their progress and development day by day—would contribute to a thorough knowledge of them and their differences.

A small Wardian case, a northern aspect, a few blocks of

sandstone, and some peat soil, are the materials that would be required for their cultivation. The Wardian case, which may be rude or polished, while protecting them in some degree from the changes of temperature incidental to a low-land climate, would secure to them a constantly moist atmosphere, which they all prefer. The interior should be fitted up with an artificial mound of "rockwork," made of lumps of porous sandstone. At the base of the "rockwork" a little pond or pool would provide a situation in which *Isoëtes* and *Pilularia* might be cultivated. In the interstices of the rockwork, the smaller and alpine species, such as *alpinum, annotinum,* and *selaginoides,* should be planted; while about its base on the margins of the water, and consequently on the lower and damper parts, should be placed such as *inundatum* and *clavatum.*

The soil employed should be peat earth intermediate in texture between the spongy and the unctuous kinds; that used among the rockwork may have in addition a portion of the sandstone pounded and intermixed with it. That used for *inundatum* in the lower part of the case will not require this intermixture, and, in fact, will be the better without it.

All parts of the soil should be kept rather moist than otherwise, by the application of fresh water occasionally; but as the confinement of the atmosphere in the damp state, in a close case, might tend to produce decay in some parts of the vegetable tissues, the little door or hinged sash may from time to time be left open for a few hours, in order that the stagnant moisture may be carried off, when a fresh supply will be doubly grateful to the plants.

It must be recollected, that the soil will be exposed to very slight drying influences, and can, therefore, never require to be very copiously supplied at any one time; the proper course being, rather to ventilate frequently, say once a week, in order to carry off the accumulated dampness, and then by a moderate fresh supply to produce a continued change of the watery element. For the same reason, and to prevent the souring of the soil, which always takes place more or less when it is in contact with stagnant water, an outlet at the bottom of the case should be carefully provided, by which all the free water at least, which drains through after the soil has been irrigated, may be removed as it accumulates.

As to aspect, the northern is decidedly the best, principally for the reason that in such a situation the sun has less influence on the temperature of the interior of the case; and an extreme degree of confined heat would be anything but favourable to these plants.

The appearance of the case would, no doubt, be improved by covering the soil entirely with living *Sphagnum* moss, which, if neatly packed on the surface with the tops of its stems uppermost, would continue to grow. Most of the species of Club-moss would prefer to grow amongst the *Sphagnum*, which, to prevent its being drawn up and smothering the plants, should be neatly clipped down occasionally with a pair of scissors.

The interest of such a collection, so far as their appearance is concerned, would depend of course upon the taste with which the rockwork was designed and executed, and the plants distributed about it; but whatever the result as a matter of taste, the study of the living plant might be prosecuted without inconvenience, and—which could never happen in their wild localities—all the species might be brought under the eye at one time, for the purpose of contrasting them, and studying their differences.

CHAPTER XXIV.

THE BRITISH PEPPERWORTS.

THE group of plants to which the name of Pepperworts has been given, is technically called *Marsileaceæ*, and contains but a few genera, these being of very curious structure. It has only two representatives in the British flora. These two plants belong to different genera, and are both submerged aquatic plants of small size, agreeing in having grassy or quill-like foliage, but differing materially in habit, the one being a creeping grower and the other tufted.

Isoëtes is sometimes classed with the Club-mosses instead of the Pepperworts. It takes its scientific name from the Greek word *isos*, equal, and *etos*, the year, from its retaining its fronds throughout the year; and is commonly called Quillwort. The genus differs from *Pilularia*, its nearest ally, in having its spore-cases enveloped by the dilated bases of its hollow leaves; some of the spore-cases containing large, and some much smaller pollen-like spores. It may also be known, when examined in a fresh state, by its hollow leaves being composed of four rows of elongated cells, which give it a bluntly quadrangular section. There is but one species, the *I. lacustris*, a stemless quill-leaved submerged plant, which gives the appearance of a green turf to the bottom of the water where it occurs.

Pilularia, the Pillwort or Pepper-grass, differs considerably from *Isoëtes* in the parts of fructification; for while in *Isoëtes* the spore-cases are within the thickened bases of the leaves; in *Pilularia* they are quite free, and attached directly to the stem, though seated at the base of a small tuft of leaves; they also differ in structure, that of *Isoëtes* consisting of granular and pulverulent spores, occupying separate spore-cases, while in that of *Pilularia* the two kinds of spores are produced within each spore-case, the larger bodies occupying principally the lower, and the smaller ones the upper parts. Its name comes from *pilula*, signifying a little pill, the spore-cases having a nearly globular form.

FIG. 28.

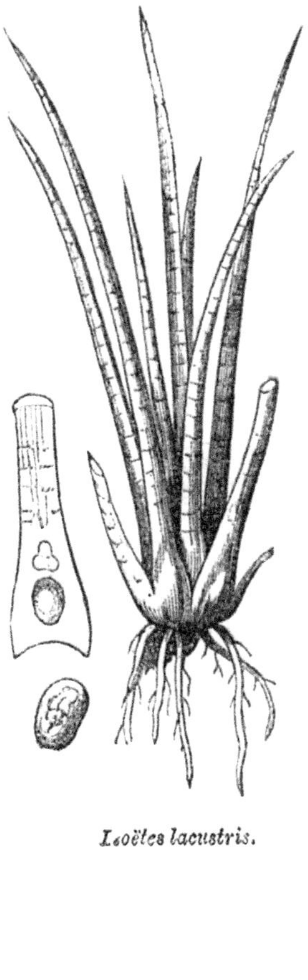

Isoëtes lacustris.

THE EUROPEAN QUILLWORT, OR MERLIN'S GRASS.

This is the *Isoëtes lacustris* of botanists, a very curious plant, growing at the bottom of mountain lakes, and having so much the appearance of submerged grass, that the inexperienced eye would probably pass it by unnoticed. It has a fleshy tuber, nearly globular in form, white and compact internally, but spongy and dark-brown coloured on the outside. The leaves spring up from the crown of these tubers, and grow erect to the height of four or six inches, or more; they are persistent, and of an olive-green colour, and their general form is awl-shaped, with the basal portion dilated; above which dilated part they are bluntly quadrangular, being formed of four parallel hollow tubes, which taper off towards the apex, and terminate in a sharp point.

The fructification is contained within the dilated bases of the leaves, and varies with the position it occupies. The spore-cases at the base of the outer leaves contain roundish bodies or spores, marked on the top by three elevated radiating ridges. The spore-cases found at the base of the inner leaves contain more numerous minute angular spores, of a pale yellow colour.

It is said that fish feed on the *Isoëtes;* and that, when brought within the reach of cattle, it is greedily eaten by them, and proves fattening.

The cultivation of the Quillwort presents few difficulties; in fact, water and a little soil are the only requisites. In such a miniature lake as has been recommended to be introduced in a Wardian case fitted up for Club-mosses, this plant and the *Pilularia* might be made to thrive; but the most interesting way in which it could be grown would be in an aquatic plant-case, with transparent sides, or in any substitute for such a structure, such as a glass jar of sufficient depth. Planted in this way, its growth could be watched, and many interesting points of its economy could not fail to reward a careful observer.

The aquatic plant-case admits of much variety of detail. The most useful form is probably that of a rectangular glass cistern of the requisite size, held together by a light metal frame, and closed in by a glass lid or cover. This would require to be supported on a stand. On the bottom of the interior, or projecting from the sides, proportionate-sized masses of coral or other rocks should be introduced, among which a little soil introduced would serve to fix and nourish the plants. Thus the smaller aquatic plants might, though in their proper element, be examined without difficulty, and at all times.

The proper situation for such a case would be the inside of any convenient window, provided it were not too much exposed to the heat of the sun; for if placed where the sun would have much influence on the temperature of the water, the plants would probably suffer. Some of the very small kinds of fish and the *small* aquatic molluscs might be introduced with advantage, and they would impart something like animation to the water. A miniature Aquarium of this kind, stocked with miniature fish, and planted with the *Vallisneria* and other aquatics in the water, and with *Trichomanes* and other Ferns above, would furnish an object of intense and ever-changing interest.

THE PILLWORT OR PEPPER-GRASS.

The *Pilularia globulifera* of botanists, called sometimes Pepper-grass, is a small creeping plant with grassy leaves, growing usually in the shallow margins of lakes and pools, where it is occasionally overflowed; but sometimes occurring entirely submerged. The stem is thread-like, occasionally branched, and producing on the lower side, at intervals, small tufts of fibrous roots which descend almost perpendicularly into the muddy soil beneath. On the upper part of the stem, at the same points, occur tufts of erect leaves, which are curled up in the incipient state, like those of a Fern, but on unrolling assume the erect position. These leaves are bristle-shaped, from one to four inches long, bright green, smooth externally, and hollow within.

FIG. 29.

Pilularia globulifera.

The fructifications consist of small globular spore-cases, attached by a very short stalk to the stem at the points whence the leaves and roots proceed. They are densely covered externally with pale brown jointed hairs, and are about the size of a small pea or pepper-corn. These spore-cases are four-celled, and when mature, split into quarters, the four parts remaining attached to the footstalk by their base. The lower part of the spore-case is occupied by the large spores, which are roundish-oblong, with a terminal nipple-like point, and the upper part is occupied by the small spores which are oblong pale yellow bodies resembling pollen.

The Pillwort is widely distributed throughout the United Kingdom, but is apparently more abundant in England and Wales, than in Scotland and Ireland. It usually grows on the margins of lakes or pools, where it is covered by the water in winter, and more or less exposed during the summer; but it is also sometimes, though rarely, met with entirely submerged.

CHAPTER XXV.

THE BRITISH HORSETAILS.

This race of plants bears an aspect altogether different from that of the foregoing groups; and indeed they have no very obvious affinity with any existing order of plants. In their mode of growth they have a certain resemblance to the *Ephedras* and *Casuarinas*, but this resemblance is confined to their general aspect. With Ferns and Club-mosses they have little in common. Their most direct relationship is with the aquatic group *Chara*.

The Horsetails are distinguished from other plants by the following characteristics. They are leafless, branching, with hollow jointed stems, separable at certain joints, which occur at intervals where they are solid, and surrounded by membranous toothed sheaths: each length, in fact, terminates above in one of these sheaths, into which the base of the next length fits. The sheaths seem to represent abortive leaves. The fructification consists of terminal cone-like heads.

The stems consist chiefly of cellular matter, coated externally by a layer of hard woody tubes, from which plates of a similar nature project towards the central cavity. Between the outer and inner surface of this cylinder-like stem, occur one or more circles of tubes, or air-cavities, differing in size and position; these afford, by their comparative size, number, and arrangement, excellent auxiliary marks for the recognition of the species. The cuticle or skin abounds in silicious particles secreted in the form of little warts, which impart to the surface a greater or less degree of roughness in proportion to their prominence. In some species this deposit of silicious matter is so great, that the whole of the vegetable substance may be destroyed by maceration, the form of the plant being preserved entire in the flinty coating. It has been found that the ashes contain half their weight of silica.

On subjecting a portion of the cuticle to the analysis of polarized light under a high magnifying power, Dr. Brewster detected a beautiful arrangement of the silicious particles, which are found to be distributed in two lines parallel to the axis of the stem, and extending over the whole surface. The greater number of the particles were seen to form simple straight lines, but the rest were grouped into oval forms, connected together like the jewels of a necklace by a chain of particles forming a sort of curvilinear quadrangle; these rows of oval combinations being arranged in pairs. Many of the particles which form these straight lines do not exceed the five-hundredth part of an inch in diameter.

Beyond their employment in the arts, the *Equisetums* are of litle importance in an economical point of view. They are useless as fodder, and exploded as physic, though they have had some reputed astringent virtues. The underground stems, however, contain in winter, when the plants are inactive, a considerable quantity of starch, and they may be occasionally eaten by animals.

The jointed tubular silicious stems, and terminal cones of fructification, are marks by which the *Equisetums* may always be readily distinguished from all other plants; but the species are not so easily recognised among themselves, owing to the great sameness which occurs among certain groups of them. The chief features relied on for their discrimination, are the similarity or otherwise of the fertile and barren stems, the number of ridges or striæ which occur on the exterior surface of these stems, and the structure of the sheaths which surround the joints. By means of the peculiarities which these parts present, the species may be certainly identified, and after a little experience has been had, several of them may be at once known by means of those first-sight appearances which become associated with the plants in the mind of the attentive student.

The name *Equisetum* is compounded from *equus*, a horse, and *seta*, a hair or bristle; whence comes the English name of Horsetail,—a not inapt comparison with the barren stems of some of the species.

THE GREAT HORSETAIL; OR, GREAT WATER HORSETAIL.

This plant, the *Equisetum Telmateia* of botanists, and called also the *Great Mud Horsetail*, is one of those species in which the ordinary fertile and the barren stems are perfectly dissimilar; the former being short and quite simple, the latter tall and compoundly branched. Occasionally a third sort of stem intermediate between the two, is produced late in the season.

The barren stems are very stately objects when in a luxuriant condition of growth. They grow erect, from six to seven feet or more in height, and are clothed nearly to the bottom with spreading proximate whorls consisting of from thirty to forty branches, which are sometimes again branched. The upper whorls have fewer branches. The whorls are most crowded towards the top of the stem, and there also the branches are about the full length—six or eight inches; lower down the stem the branches become shorter, and the whorls more distant. The stems measure about an inch and a half in diameter at the stoutest part, and from this point decrease upwards, becoming slender at the point. The surface is smooth, with mere indications of about thirty faint lines extending into the sheaths, and there becoming more apparent. The sheaths set close to the stem, or nearly so, and are half an inch long, green below, with a dark-brown ring at top, and divided at the margin into slender, bristly, dark-brown teeth, with paler membranous edges, and frequently adhering together in twos and threes. The branches have eight or ten ribs united in pairs, and their sheaths terminate in four or five teeth.

The fertile stem is erect, simple, from nine inches to a foot or more high, succulent, pale brown, and smooth. From each of the numerous joints arises a large loose funnel-shaped sheath, the upper ones being largest; they are distinctly striated, and terminate in thirty to forty long, slender teeth. The catkins are large, between two and three inches long.

A section of the barren stem shows an outer surface without ridges and furrows, and in the very narrow cylinder of the stem occur two circles of cavities, the outer one consisting of large openings, those of the inner minute, and alternating with the larger. The central cavity is very large, the tissue of the stem being reduced to a very narrow ring.

This is a widely-dispersed and rather common plant, occurring on moist banks and in muddy places, by the sides of streams and the margins of muddy pools. The nature of the soil would seem to be of small importance, provided it has its necessary degree of moisture, for it is recorded as occurring both in sandy and in clayey soils, as well as in muddy pools. It is frequent in Ireland; and is found both in Scotland and Wales.

THE SHADE HORSETAIL.

This plant is the *Equisetum pratense* of botanists; and has been also known in this country as *E. umbrosum*, and *E. Drummondii.*

The fertile and barren stems are quite dissimilar in their appearance. The former are short, quite simple, and terminating in a cone-like head of spore-cases. The latter are taller, and produce several whorls of long, crowded, slender branches; whilst a third kind produce both whorls of branches and cones. In the production of these three kinds of stems it serves to connect, through *E. sylvaticum*, that group in which the fertile and barren stems are successive and altogether unlike, with that in which the stems indifferently bear the fructification.

The fertile stems grow about six inches high, and are quite branchless; they have numerous joints, the large loose funnel-shaped pale-coloured sheaths produced at these points, often almost covering the stem. The teeth, which terminate the sheaths, are awl-shaped, pale brown, with pale-coloured membranous margins, and number from twelve to twenty, equalling the ribs. The fructification forms a moderate-sized, terminal, oval, cone-like head.

The barren stems grow erect, eighteen inches or more in height, and have on their surface about twenty sharp ridges, with corresponding furrows, the ridges being coated with prominent silicious warty particles, so that the stems are very rough. The few lower joints are without branches, but those in the upper part of the stem produce whorls of from ten to sixteen branches, which are simple, and at first drooping, but eventually become spreading. The sheaths of these barren stems are much smaller than those of the fertile, less funnel-shaped, and more closely set to the stem, and their teeth are also fewer, shorter, and blunter. The branches are slender, three or four-ribbed, and have loose sheaths, which terminate in short, acute, membranous-edged teeth.

The branched fertile stems have their sheaths smaller than the simple fertile ones, but larger than the barren ones. Several of the uppermost joints produce whorls of branches, and the stem is terminated by a cone of fructification. In these cases, however, the number of branches is less than that produced by the ordinary barren stems, and the cone is smaller than those produced by the ordinary fertile stems.

The section of the stem shows on the exterior a series of sharp ridges with angular furrows; the central cavity rather exceeds a third of the whole diameter; the cylinder of the stem is then pierced by three circles of cavities—one of longish oblong openings opposite the furrows, one of minute pores exterior to these and opposite the ridges, and another of minute pores on their inner side also opposite the ridges.

Probably this species is tolerably plentiful in moist shady woods, which are the situations it affects; but it has as yet been met with only in a limited number of localities in Ireland, Scotland, and the north of England.

THE CORN-FIELD HORSETAIL.

This is the *Equisetum arvense* of botanists. It is the most common of the species, and in many places is an injurious weed, very difficult to eradicate. It occurs here and there, almost everywhere, in fields and waste places, especially where the soil is sandy. It has long, creeping, underground stems, which are a good deal branched, and are cylindrical and jointed in the same way as the stems which rise aboveground. The stems which appear aboveground are of two kinds, the one simple and fertile, the other branched and barren.

The fertile stems are quite without branches, and grow up early in spring, in April and May, arriving at maturity and perishing long before the barren ones have completed their growth. They are from three to eight or ten inches in height, hollow, succulent, and nearly smooth. The sheaths are large and loose, widening upwards, pale-coloured, divided into about ten dark-brown teeth, which often adhere together in twos and threes. The teeth are very narrowly lance-shaped and sharp-pointed, and correspond with the ribs, about ten in number, by which the sheaths are marked. These stems are terminated an inch or two above the upper sheath, by cone-like heads, rather more than an inch long.

The barren stems are either erect or decumbent, and from one to two feet or more in height; they are generally branched from bottom to top. They spring up after the fertile stems have withered, and are at first crowded with short appressed branches, which, by degrees, become elongated, and spreading, and are sometimes again branched. The main stem has from ten to sixteen distinct shallow furrows, with corresponding ridges, and is, as well as the branches, studded over with minute silicious warty particles. The sheaths, which fit somewhat closely to the stem, are furrowed like it, and terminate in an equal number of acute wedge-shaped dark-coloured teeth, which are often margined by a narrow brown membrane. The branches are four-ribbed and four-angled, and their sheaths four-toothed, the teeth being long and acute.

The section of the stem of *E. arvense* shows an interior cavity occupying only about one-third of the diameter. The exterior surface is varied by about a dozen blunt ridges, having corresponding shallow depressions; within this, occupying about the centre of the ring, and alternating with the ridges, are a series of large roundish-oblong or obovate cavities, the narrow end of which is turned inwards; alternating again with them, and consequently opposite to the external ridges, occurs an annular series of small circular cavities, which are placed near the inner surface of the tube.

This plant is not applied to any use; and the harshness of its stems renders it by no means agreeable to cattle, although, in some situations, it occurs abundantly among their pasturage; and in cultivated ground becomes a troublesome weed.

THE WOOD HORSETAIL.

This species is the *Equisetum sylvaticum* of botanists. It is perhaps the most beautiful of the *Equisetums;* certainly it is extremely elegant in almost all stages of its growth, and perhaps never more so than shortly after the fertile stems, with their fructification still perfect, have begun to develop their lateral branches. Later in the season these branches, which have from the first a pendent tendency, droop around with exquisite grace on all sides.

FIG. 30.

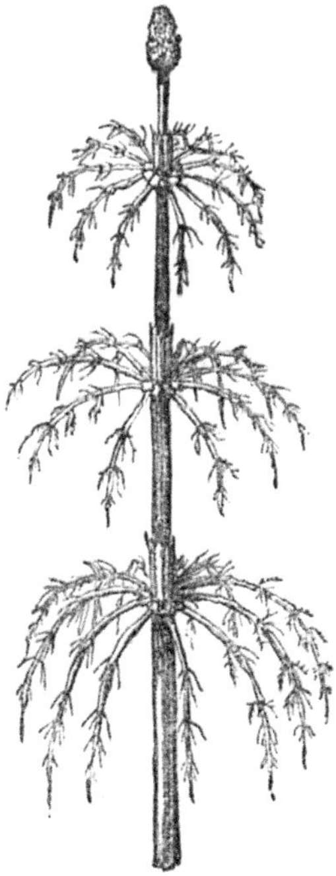

Equisetum sylvaticum.

The stems are erect, and in a certain sense, those of them which produce fructification, and those which are barren, are similar, except as regards this one point. Their resemblance consists in both growing up at the same time, and both putting out whorls of deflexed branches, which are, however, less numerous on the fertile stems. In other respects they differ, as, for instance, in the growth of the apices of the fronds; for the fertile ones, terminating in a catkin which soon perishes, become blunt-topped, while the barren ones continue to elongate at the point and so become pyramidal. The barren stems are also more slender than the fertile ones, and have less inflated sheaths. This species, therefore, in its habit of growth, holds a middle place between that group in which the fertile and barren stems are successive and quite dissimilar, and that group in which they are simultaneous and present no appreciable difference of structure.

The fertile stems, when they first shoot up, are almost quite simple, and a few of them remain so, perfecting their cone-like head, and then perishing. More usually, by the time the catkin has become fully grown, the whorls of branches from the upper joints will be seen protruded to the length of from half an inch to an inch or rather more. Two, three, or four, rarely more, whorls of branches are thus produced from the uppermost joints of the stem, and above these the oblong-ovate blunt cone is seated on a bare stalk-like portion of the stem. The stems are about a foot high, round, succulent, pale-coloured, with about twelve slender ridges and corresponding shallow furrows, nearly smooth, the silicious particles which coat the surface being too minute to impart much roughness; they terminate in an oblong, ovate cone of fructification. The sheaths are large and loose, and are divided at the margin into three or four bluntish lobes; their lower half or tubular portion is pale green, the upper half or lobes bright russet. The slender branches, which are deflexed, grow to about a couple of inches in length, and produce from their joints a series of secondary branches, which grow from about half an inch to an inch.

The barren stems are taller, more slender, and less succulent, and also produce more numerous whorls of branches. The sheaths fit closer than those of the fertile stems; and the whorls of branches are very dense, being compoundly branched. The main branches are three-ribbed, their joints terminating in three long pointed teeth; they are about four inches in length, constantly branched at every joint with a whorl of branchlets averaging two inches in length, and sometimes these branchlets put out another series of shorter branchlets. The outline would be pyramidal, were it not that the extreme point becomes so slender as to be unable to retain itself erect. The lateral branches are all drooping or deflexed, and hence the elegant appearance of the full-grown plants.

The section of the stem shows a series of shallow ridges and furrows; opposite the latter a ring of largish cavities; and alternating with these on the inner side, another ring of very minute cavities, these latter again alternating with a circle of angular cavities close to the inner margin of the tube. The central cavity measures about half the diameter.

This species grows naturally in moist shady woods; and though local, owing apparently to the conditions necessary to its growth, namely, shade and moisture combined in a peculiar way, it is, nevertheless, a widely distributed plant, and can hardly be considered as uncommon throughout the United Kingdom. Its fertile stems are in perfection about the middle of April, and its barren stems in June.

THE WATER HORSETAIL.

This, the *Equisetum limosum* of botanists, is sometimes called the *Smooth Naked Horsetail.* It is a common species and generally distributed, occurring principally in pools, ditches, and marshy places, though occasionally in running streams. It is rather tall-growing, the stems rising from two to three feet or more in height; these, though finely ribbed, are smooth to the touch, the furrows being very shallow; their smoothness no doubt arising from the coating of silicious particles being much finer and less prominent than in others which are more harsh to the touch. Sometimes the stems are quite unbranched, sometimes furnished with irregular whorls of branches along all their central portion; and between these two extremes there occurs every conceivable degree of branching, from the single shoot produced here and there, through every gradation of imperfect whorls up to whorls of short branches almost complete. The branches, which are simple, nearly erect, and never acquire much length, are from four-angled to eight-angled, and are smooth like the stem. There is no material difference between the barren and fertile stems, except the presence of the fructification in the one case and not in the other.

The surface of the stem is marked with from sixteen to twenty very slight ridges, and the sheaths, which are short, rather closely fitted to the stem, and of the same colour in the lower part, terminate in an equal number of dark-coloured awl-shaped teeth, which sometimes have a pale membranous margin.

Owing to the shallowness of the ridges and furrows, the section of the stem shows a nearly smooth exterior outline, and the cylinder of the stem is furnished only with a row of minute cavities near the inner margin; this cylinder is very thin compared with the diameter of the stem, the central cavity being unusually large.

This plant is the most fodder-like of any of the *Equisetums*, owing to its less flinty cuticle, but in this point of view, it is, at least in this country, of very small importance. It is, however, stated to be used in Sweden as food for cattle, "in order that the cows may give more milk;" and in Lapland, it is, even when dry, eaten with avidity by the reindeer, though they will not touch common hay. Linnæus censures the improvidence of the Laplanders, in not providing during summer a supply of this plant and of the Reindeer Moss, for winter use; thus making some provision for their herds at a time when the ground is covered with frost-bound snow, so as not to risk the loss of their most valuable or entire possessions. An instance is related by Mr. Knapp, in which a colony of the short-tailed water-rats made this plant their food, and in the evening might be heard champing it at many yards' distance.

THE MARSH HORSETAIL.

The *Equisetum palustre* of botanists. A common species in boggy places and by the sides of ditches and water-courses. The stems are erect, growing from a foot to a foot and a half in height, the presence of fructification alone distinguishing the fertile from the barren. They are somewhat rough on the surface, but less so than in many other kinds; and they are marked on the exterior by prominent ribs, with intervening broad deep furrows, the number being variable, from six to eight. The joints are invested with nearly cylindrical sheaths, which are quite loose, and in the upper parts of the plant almost twice the diameter of the stem. The sheaths terminate in as many acute wedge-shaped pale-coloured teeth as there are ridges on the stem. The stems are usually, except at the base, furnished with whorls of numerous simple branches, the number corresponding with the furrows of the stem; these are slender, four or five-ribbed, and their sheaths set nearly close.

A section of the stem shows a series of prominent ridges on the outer face; just within these, and over against the furrows occur a circle of moderate-sized cavities: and alternating with these, and near the inner margin, is a series of much smaller circular cavities. The central cavity of the stem is comparatively very small, not much larger than the series of openings near the outer surface. The resemblance is considerable between its section and that of *E. arvense*.

Besides the foregoing usual form, there are some curious variations to which this plant is liable. One of the most remarkable has been called *polystachyon*, and is remarkable in having more or less of the branches of the two upper whorls terminating in cones of fructification; the usual habit of the plant being to produce only one cone, and that on the central stem. It has been suggested that the production of these lateral fructifications is accidental, owing to the destruction of the top of the main stem, but this explanation is quite insufficient, since they are sometimes produced along with the central head, which moreover varies when accompanied by them, being sometimes of the usual size, and sometimes reduced in size like the lateral heads. The lateral heads are usually later in their appearance than the central ones. Occasionally some of the branches of the lowest whorl become elongated, and terminate in these small cones.

Another form, called *nudum* or *alpinum*, is a depauperated form, differing from the ordinary plant in being altogether smaller, the height ranging from two to four or five inches, the lower part of the stems being decumbent, and the whole stem almost devoid of branches; a few being developed only at their very base. In some states, this form has much resemblance to the prostrate *E. variegatum*, but is distinguishable by means of its sheaths and fructification.

THE LONG ROUGH HORSETAIL.

This species is by botanists called *E. ramosum*, and has been in this country also known as *E. elongatum* and *E. Mackayi.* It is one of those species in which the stems that produce the fructification, and those which are barren, do not differ in any other respect, and are therefore said to be similar. The stems also are almost branchless, the branching being mostly confined to the production of one or two erect lateral stems from near the base, and this lateral branching is by no means common. Sometimes, indeed, the upper part of the stem is also sparingly branched, but the branches are produced singly from the joints. The stems are slender and erect, from two to three or four feet high, deeply furrowed, with a double row of elevated points along the ridges, which are usually from eight to twelve, but sometimes fourteen in number. The sheaths are close, cylindrical, and striated like the stem, terminating in a number of teeth equalling the striæ; these teeth are long, slender, awl-shaped, black with pale membranous margins, and usually, but not always, persistent. The sheaths are, for the most part, entirely black, but here and there they occur with a narrow greyish ring.

The section of the stem differs from that of *E. hyemale*, to which it presents a general resemblance, in being smaller, showing fewer ridges, and having the cavities placed rather nearer the inner margin; the central cavity is also proportionally smaller. It has, consequently, on the exterior, a series of ridges formed of twin projections representing the double row of silicious particles which extends along each ridge; and a series of cavities rather nearer the inner than the exterior surface of the ring.

This plant has been found on the moist banks of the mountain glens of Scotland and the north of Ireland.

THE GREAT ROUGH HORSETAIL.

This plant is the *Equisetum hyemale* of botanists. The items are of a deep glaucous green, and all alike in structure, those which bear fructification differing in no other particular from those which do not. They grow upright, from two to three feet high, and are scarcely ever branched: when this does occur a solitary branch is produced, and this protrudes from below the base of one of the sheaths of the stem; they are cylindrical, tapering off at the apex, and marked on the thicker parts with from fourteen to twenty ridges, formed of a double row of elevated points, consisting of crystallized silicious particles; hence the stems are very rough. In this species the sheaths fit closely around the stems, so that they are nearly cylindrical; they are marked by the same number of ridges as the stem, but they are less prominent, and terminate in a series of black, membranous, bristle-shaped teeth, which soon fall off, and leave the margin crenated. The sheaths are at first pale green with a black margin; from this they change to be entirely black; and finally they become whitish in the middle, leaving a narrow ring of black at the base and margin.

In this species a section of the stem shows on the exterior a series of distinct ridges, formed of twin projections, and varying in number, as has been already explained; opposite to the furrows, between them, and occupying about the centre of the solid cylinder, is a ring of moderate-sized cavities. The central cavity is comparatively large.

This plant grows naturally in boggy shady places, and is much more abundant northwards than southwards, where it is rarely met with. Though distributed sparingly over the United Kingdom, its occurrence is strictly local.

The stems of this *Equisetum* are employed in the arts as a material for polishing, and are imported under the names of

Dutch Rush and Shave-grass. They are obtained from Holland, where this species is planted to support the embankments, which it does by means of its branching underground stems. It has been suggested that our own sandy sea-coasts might be profitably planted with it. The peculiarity which gives it its commercial value, is the presence of a very hard coating of silex, which is deposited in the form of little crystals, rendering the surface rough like a rasp or file, and hence not only woods, but metals and stones may be polished by it. This silicious coating is so entire, and of such density, that it is stated the whole of the vegetable matter may be removed by maceration, or, according to others, by burning, without destroying the form of the plant. The minute crystals of silex, of which the flinty coating consists, are arranged with a degree of regularity which, under a microscope, has a very beautiful appearance; they form a series of longitudinal elevated points, and in the furrows between them are cup-shaped depressions, at the bottom of each of which is placed a stomate or pore.

All the species of *Equisetum* have a flinty coating to their stems, and may be, and are, more or less employed in polishing; but the stems of the *E. hyemale* are much preferable to those of the other kinds, in consequence of their rougher and more hardened surface.

MR. MOORE'S ROUGH HORSETAIL.

This is the *Equisetum Moorei* of Newman. It differs from the other native unbranched Horsetails in the nature of its stems, which are not persistent through the winter, or evergreen, as they are, but die down in autumn, and are renewed; they are therefore annual. They grow a foot and a half to two and a half feet high, and are unbranched, except where the apex has been destroyed, in which case branches are sparingly produced. They are rough, and are channelled with about twelve deep well-marked furrows. The sheaths, which are loose, and have the same number of ridges as the stem, are whitish, with a black ring at the base and tipped by about twelve blackish teeth, which are rigid, bluntish, and terminated by elongated membranous paler awns.

This plant was found growing on banks facing the sea at Rockfield in the county of Wicklow, in the year 1851, by Mr. D. Moore, Curator of the Royal Botanic Gardens at Glasnevin, Dublin.

THE VARIEGATED ROUGH HORSETAIL.

This species, the *Equisetum variegatum* of botanists, is found on the banks of rivers and lakes, and in sandy places near the sea. It is one of the species whose stems are all similar, and almost quite unbranched. It extends by means of a widely creeping underground stem, producing numerous above-ground stems, often springing from joints in such close proximity, that they appear in dense tufts. Though so numerously branched just beneath or at the surface of the soil, branches are seldom produced on the exposed part of the stems; but when this does occur, they spring singly from the joints, and have much similarity to the stem itself. The stem grows about a foot high; its surface is very rough, and impressed with from four to ten furrows, alternating with rather prominent ridges, each ridge margined on both sides, with a line of minute silicious points, which give it the appearance of being grooved, and impart to it its peculiar roughness. The sheaths are slightly enlarged towards their margin, ribbed like the stem, green in the lower part, black above, and terminate in a fringe of black teeth, equalling the ribs in number, with a broad white membranous border, in form ovate, and tipped by a deciduous bristle. A certain number of the stems, usually the most vigorous, terminate in a cone of fructification. This is small, elliptic, crowned by a prominent point or apiculus.

The section of the stem shows a small central cavity, an exterior surface of rather prominent ridges, each channelled so as to form two projecting angles, and a circle of moderate-sized cavities occurring about the centre of the tissues.

A variety of this, sometimes called *E. arenarium*, is smaller and more slender, its stems always procumbent, and not having more than six furrows.

Another variety is the *E. Wilsoni*, which is a stouter and taller plant, three feet high. The section of its stem shows the central cavity and the ring of cavities occurring in the cylinder of the stem much larger than in the ordinary forms of *E. variegatum*. This plant grows in water at Mucruss, in the immediate vicinity of the Lakes of Killarney. The stems are tufted, generally simple, but sometimes sparingly branched; they have about ten furrows, with broad intermediate ridges, on which the silicious particles are less prominent, so that the stems are not nearly so rough as in the allied *E. variegatum*, *Mackayi*, etc. The sheaths are scarcely larger than the stem, and are entirely green, except a narrow, black, sinuous ring at the margin; the teeth are short, generally blunt, and have obscure membranous margins, and deciduous awns.

The present species is rather a local plant, but is widely dispersed in the three kingdoms, the larger forms growing on the margins of lakes, canals, rivers, ditches, etc., the smaller prostrate examples occurring on the sandy sea-coasts.

The *Equisetums* appear to submit readily to cultivation. The plan is to pot them in loamy soil, and to place the pots in a cold frame, among a collection of hardy Ferns; or, in the case of the aquatic species, to sink the pots just beneath the surface of a tank of water.

There are, it should be remarked, two sets of *Equisetums*, which may be called the evergreen and the deciduous groups; the former consisting of *E. hyemale, ramosum*, and *variegatum;* the latter including all the remaining species which die down in autumn, and are renewed in spring.

The evergreen species are desirable plants for damp shady rockwork, requiring no especial care or culture. Their peculiar form and character render them interesting plants, no less for their own sakes, than for the effect which their distinct appearance may help to bring out in such situations.

The most desirable of the deciduous kinds for the garden are *E. Telmateia*, *E. sylvaticum*, and *E. umbrosum*, these being the most elegant of the race. They require shade, but nothing else beyond what well-constructed rockwork would supply.

Perhaps the most interesting way of cultivating these plants would be as a separate group on a shady border. In damp cool soil they would be certain to succeed. The smaller delicate sorts, such as the procumbent *E. variegatum*, should be rather elevated between three or four rough stones, over which they would spread; and for the aquatic species, earthenware pans might be sunk, and these, half-filled with mud, and the remainder with water, would provide all that would be necessary for their well-being. All the other species would grow in the ordinary soil, provided it were sufficiently moist and cool in summer; but the rambling propensities of the underground stems should be checked by planting them in pots sunk in the ground.

The raising of the *Equisetums* from the spores, too, would be very interesting employment, and withal very instructive. The spores are very curious bodies, of a somewhat oval form, having four elastic threads, thickened at the ends, coiled around them. These, when the spore has become ripe, unroll; and their elasticity, no doubt, contributes to burst the case in which the spores are contained, as well as to assist in the dispersion of these minute reproductive bodies. They are, indeed, so irritable, that a change of temperature or moisture, such as that produced by breathing on the spores, is sufficient to produce this forcible uncoiling. The spores themselves are very interesting microscopic objects.

The germination of the spores has been made the subject of experiment by several inquirers, whose observations have been published. It appears that from three to fourteen days after the spores are sown, they send down a thread-like transparent root somewhat thickened at the end, and protrude a confervoid, cylindrical, obtuse, articulated thread, which is either two-lobed or simple at the apex. Some days after this, several branches are produced, and become agglutinated together, forming a body resembling a bundle of confervoid threads, each of which pushes out its own root. These confervoid threads go on growing and combining until a considerable cellular mass is formed. Then this mode of development ceases, and a young bud is formed, which produces the stem of an *Equisetum*, at once completely organized, with its air-cells, its central cavity, and its sheaths, the first of which is formed before the elongation of the stem out of the original cellular matter.

To watch the minute atoms thus springing into life, developing by degrees their tiny stems, and gaining strength and bulk day by day, until they reach maturity, could hardly fail, one would think, to lead a sensitive mind to pure and wholesome thought,—calling up, on the one hand, the contemplation of the wise and beneficent plans and the all-sufficient power of the Creator, by whose ordaining providence life interminably renewable had thus been made to spring from the dust-like spore; and at the same time producing, on the other, a just appreciation of the uncertainty and insufficiency of human agency. For, though man may plant and water, yet it is God alone that giveth the increase.

INDEX

OF SCIENTIFIC AND ENGLISH NAMES.

*** The vowels marked, are those upon which emphasis is to be placed in pronouncing the names. When the short accent (′) is used, the vowel is to be sounded in conjunction with the following consonant. When the long accent (ˋ) is used, the vowel is to have an open sound as though standing alone. Non-classical readers are reminded that every vowel represents a syllable in Latin or Latinized words: thus *septéntrionàle* is not spoken sep-ten-tri-o-nale, but sep-ten-tri-o-na-le; and *polypodioides* is not po-ly-po-di-oides, but po-ly-po-di-o-i-dès.

THE END.

I.

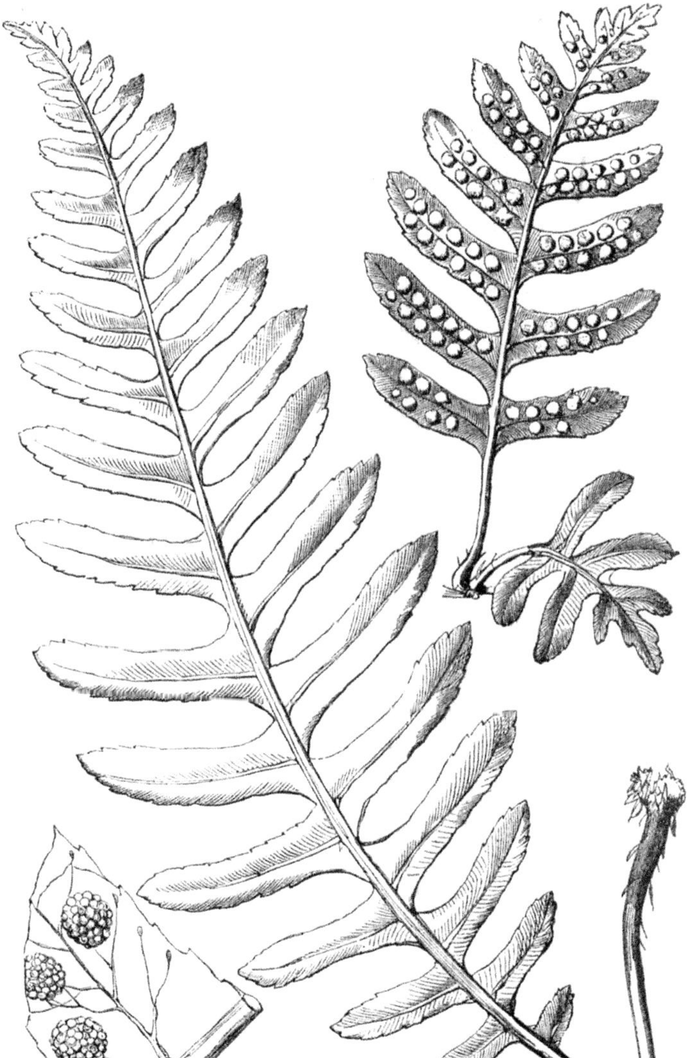

Polypodium vulgare.

II.

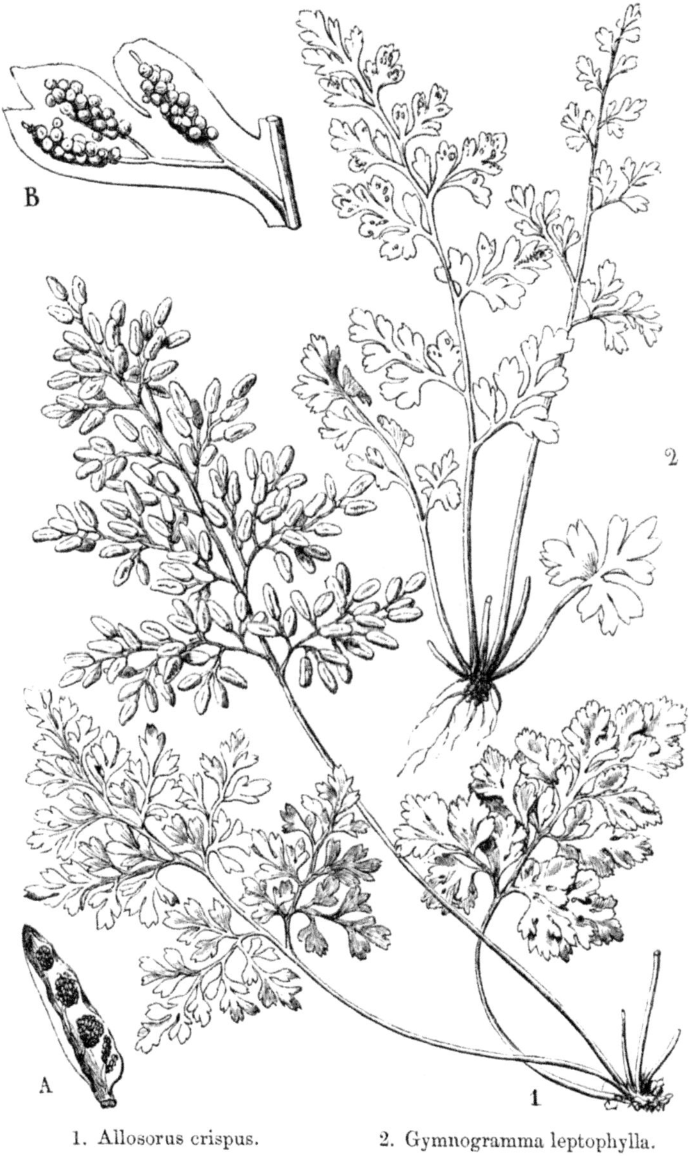

1. Allosorus crispus. 2. Gymnogramma leptophylla.

III.

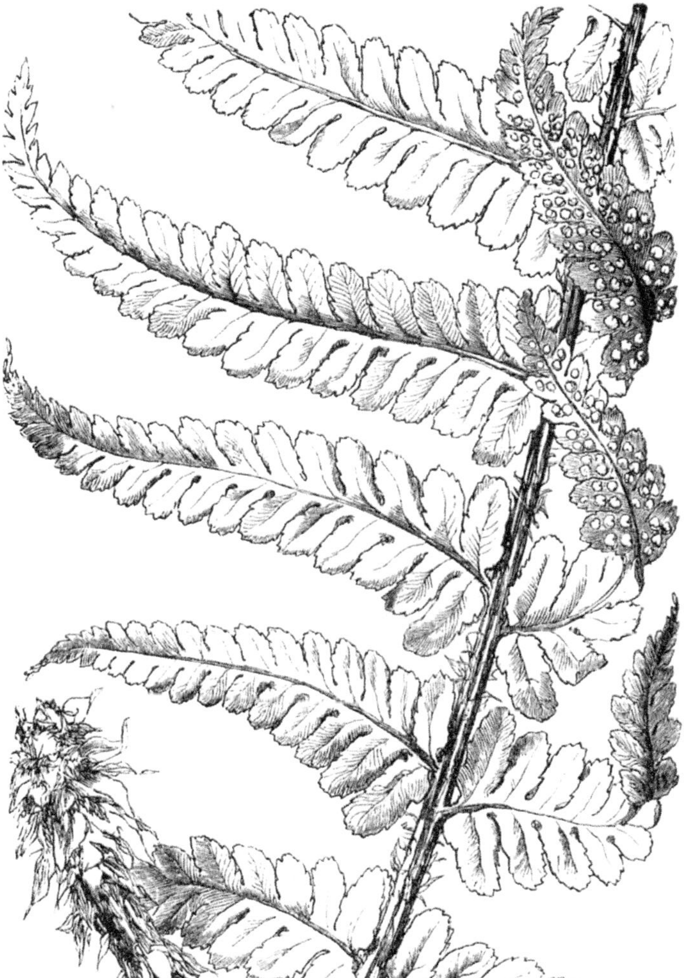

Lastrea Filix-mas.

IV.

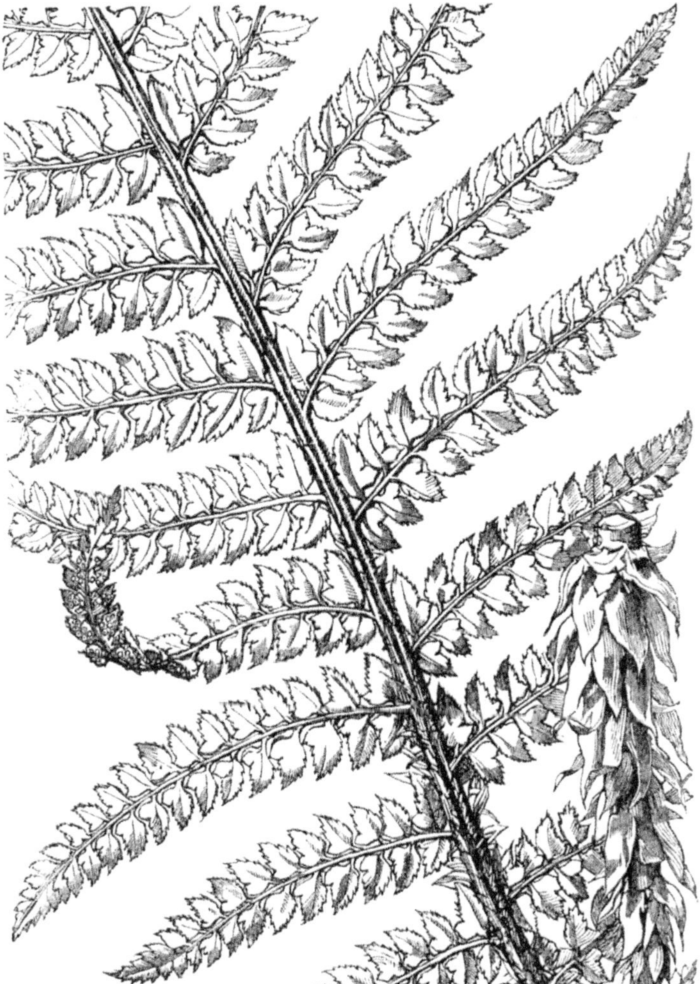

Polystichum angulare.

V.

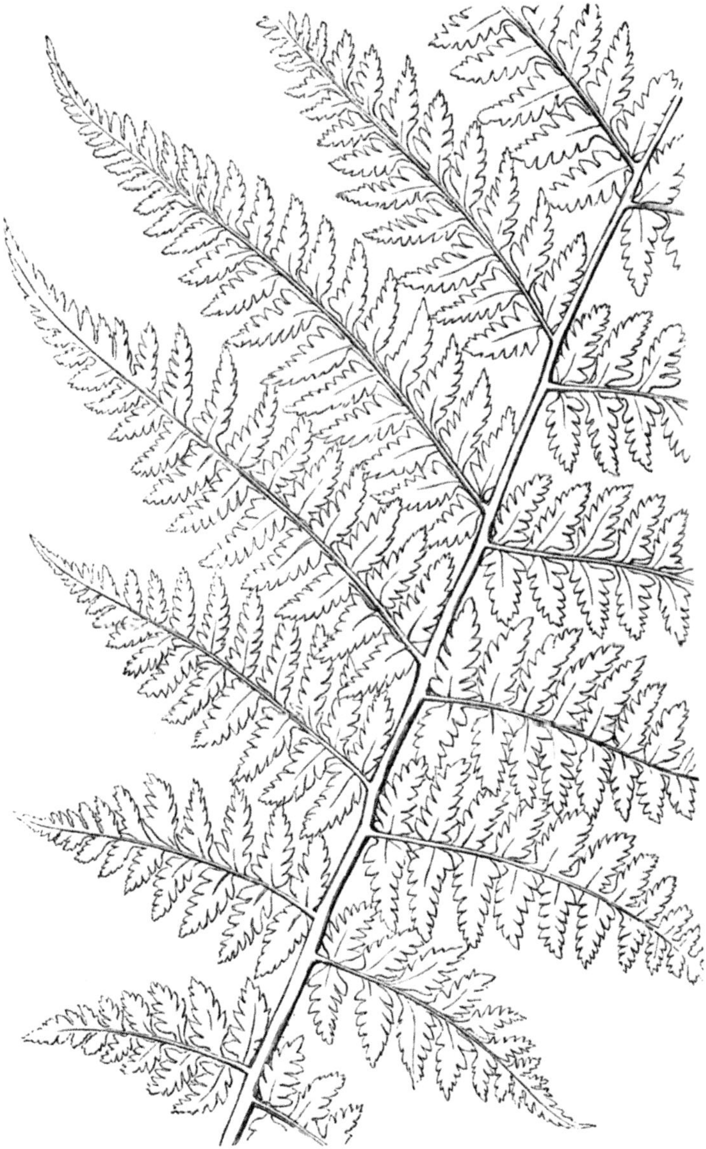

Athyrium Filix-fœmina.

VI.

1. Asplenium marinum. 2. Asplenium Ruta-muraria.
3. Ceterach officinarum.

VII.

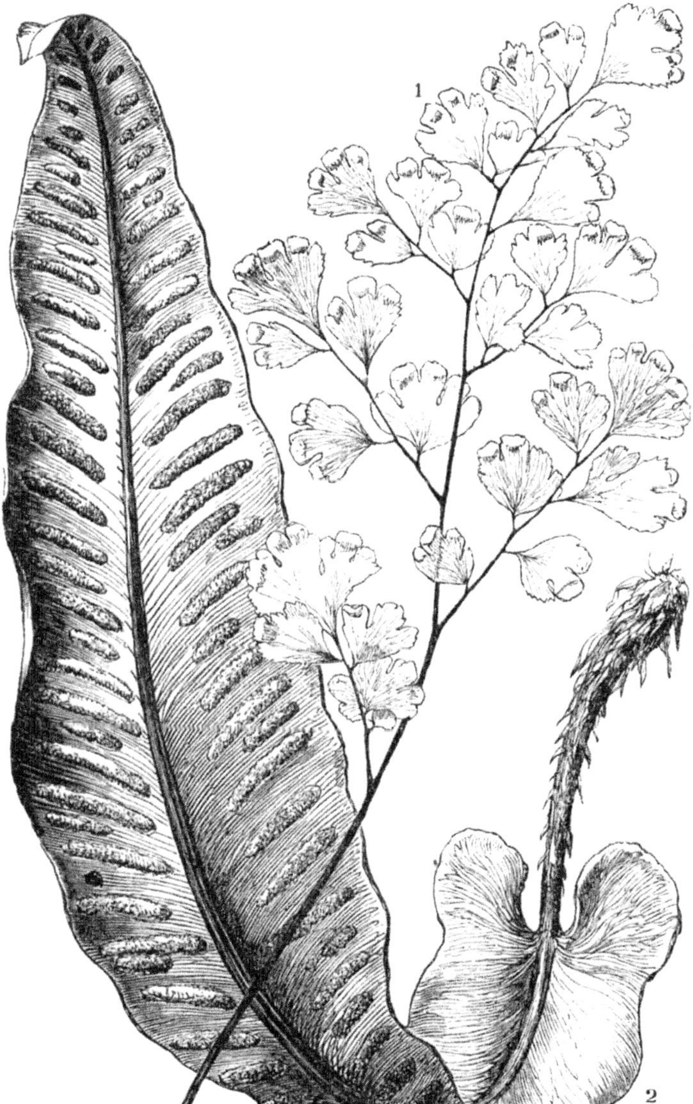

1. Adiantum Capillus-veneris.

2. Scolopendrium vulgare.

VIII.

1. Blechnum Spicant. 2. Woodsia ilvensis.

IX.

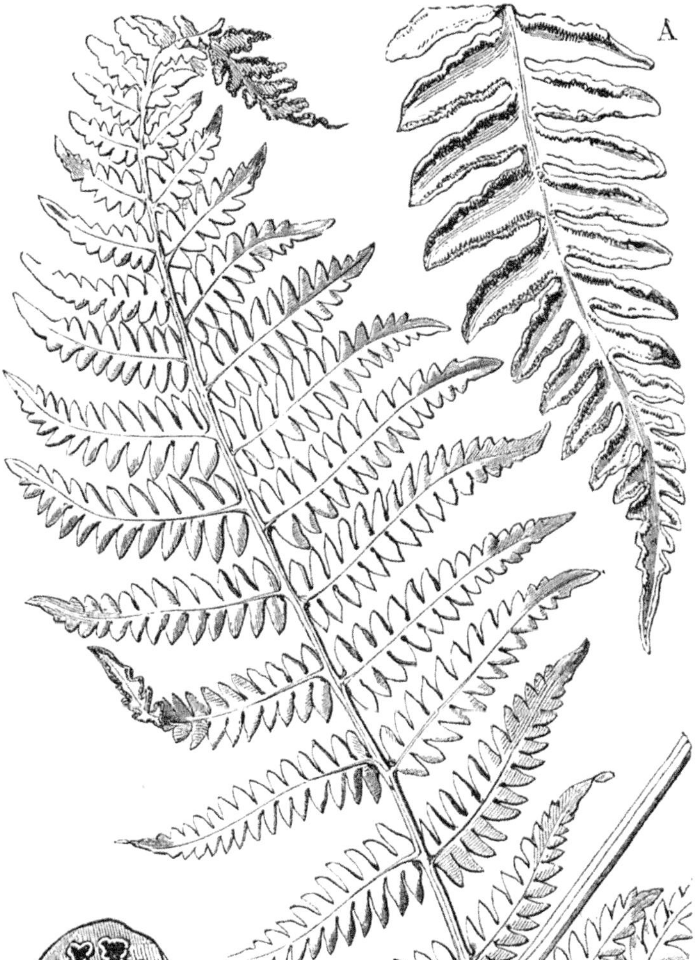

Pteris aquilina.

X.

1. Trichomanes radicans. 2. Cystopteris fragilis.

XI.

Osmunda regalis.

XII.

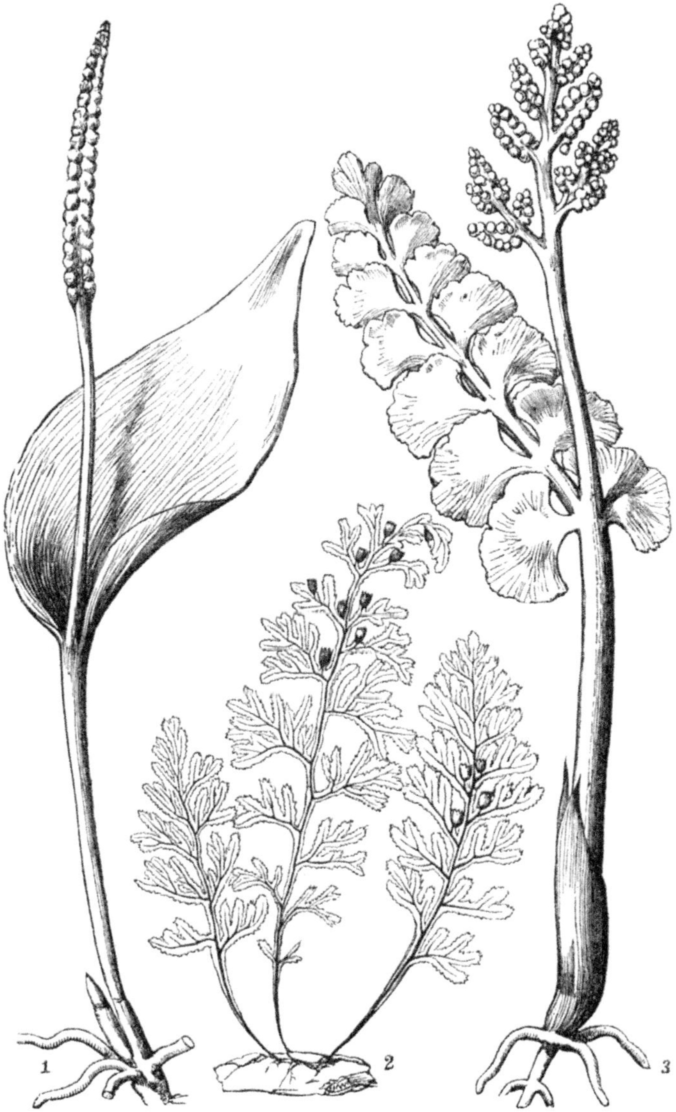

1. Ophioglossum vulgatum. 2. Hymenophyllum tunbridgense.
3. Botrychium Lunaria.

Price 3s. 6d. each, cloth gilt, or 4s. gilt edges.

SWISS FAMILY ROBINSON: or, Adventures on a Desert Island. Complete Edition, Revised and Improved. With Sixteen large Engravings by John Gilbert.

"The purport of this pleasing story is to convey instruction in the Arts and Natural History, and at the same time to inculcate by example principles which tend to the promotion of social happiness. It well deserves the high encomium passed on it by one of the most competent judges, that it is one of the best children's books ever written."

THE DAWNINGS OF GENIUS, exemplified and exhibited in the Early Lives of Distinguished Men. By the Rev. T. A. Buckley, M.A., Oxon, with Illustrations, Views, and Designs by Godwin.

The characters noted are—Pico di Mirandola, Petrarch, Giotto, Michael-Angelo Buonarroti, Raffaello d'Urbino Sanzio, Erasmus, Sir Thomas More, King Edward the Sixth, Richard Evelyn, William Wotton, Barretier, Candiac de Montcalm, Pascal, Sir Christopher Wren, Grotius, Scaliger, Gaspar Barthius, Isaac Casaubon, Sir Francis Bacon, Sir Isaac Newton, Galileo, Mozart, Niebuhr, Dr. Arnold, Hogarth, Sir Thomas Lawrence, Chatterton, Walter Scott, Franklin, Washington.

VOYAGE AND VENTURE; or, Narratives of Perils by Sea and Land. With Eight Illustrations.

"A volume of excitement for the fireside; in it are narrations of perilous incidents in naval life—by famine, fire, and wreck, as well as in the busy stir of war; the calm decision of Nelson at the Nile; the career of the soldier in tented field and siege; an instance of the firm and steady purpose of Wellington at Vittoria, as well as his kindly care for others, by an eye-witness. Here, also, are the adventures of hardy hunters in the Indian wilds, and their many hair-breadth escapes when in pursuit of savage animals, or their greater peril, the wild Indian."

TRAVELS OF ROLANDO (THE); or, A Tour Round the World. By Lucy Aiken. A New Edition, revised and corrected by Cecil Hartley, A.M. Illustrated by William Harvey.

"Combined with the most vivid descriptions, we find in the 'Travels of Rolando' the liveliest, most varied, most extraordinary, and most impressive personal adventures that imagination is capable of conceiving, or pencil of portraying."

ROMANCE OF ADVENTURE (THE); or, True Tales of Enterprise. Second Edition, enlarged and revised. With Illustrations by J. Gilbert and W. Harvey.

"Books of personal adventures are, after all, the greatest favourites with the young; and such books as the 'Romance of Adventure,' containing much information, as well as amusing and useful facts in Natural History entertainingly conveyed, cannot, we feel convinced, in the estimation of any intelligent boy or girl, fail of becoming a great favourite."

Printed in Dunstable, United Kingdom

91218258R00090